STUDY GUIDE AND MAPPING WORKBOOK

Vincent J. Del Casino, Jr.
California State University, Long Beach

World Regions in Global Context

People, Places, and Environments

Fourth Edition

MARSTON ▪ KNOX ▪ LIVERMAN ▪ DEL CASINO ▪ ROBBINS

Prentice Hall

Boston Columbus Indianapolis New York San Francisco Upper Saddle River
Amsterdam Cape Town Dubai London Madrid Milan Munich Paris Montréal Toronto
Delhi Mexico City São Paulo Sydney Hong Kong Seoul Singapore Taipei Tokyo

Geography Editor: Christian Botting
Editor in Chief, Geosciences and Chemistry: Nicole Folchetti
Marketing Manager: Maureen McLaughlin
Assistant Editor: Jennifer Aranda
Managing Editor, Geosciences and Chemistry: Gina M. Cheselka
Project Manager: Traci Douglas
Operations Specialist: Amanda A. Smith
Supplement Cover Designer: Paul Gourhan
Cover Image Cerdit: Crescent Moon Lake Oasis in Gobi Desert near Dunhuang, China © Jose Fuste Raga / Cor

Pearson Prentice Hall
Upper Saddle River, NJ 07458

Printed in the United States of America

10 9 8 7 6 5 4 3 2 1

ISBN-13: 978-0-321-66775-5
ISBN-10: 0-321-66775-1

Prentice Hall
is an imprint of

PEARSON

www.pearsonhighered.com

Table of Contents

Preface: Introduction to Mapping v
CHAPTER 1 World Regions in Global Context 1
CHAPTER 2 Europe 13
CHAPTER 3 The Russian Federation, Central Asia, and the Transcaucasus 25
CHAPTER 4 Middle East and North Africa 37
CHAPTER 5 Sub-Saharan Africa 49
CHAPTER 6 The United States and Canada 62
CHAPTER 7 Latin America and the Caribbean 75
CHAPTER 8 East Asia 87
CHAPTER 9 South Asia 101
CHAPTER 10 Southeast Asia 114
CHAPTER 11 Australia, New Zealand, and the South Pacific 126
Appendix A Answers 139
Appendix B Population Reference Bureau 2009 World Population Data Sheet 169

Preface: Introduction to Mapping

Introduction

This workbook provides you with basic map reading skills and offers you exercises to further analyze the spatial dimensions of global and regional environmental, political-economic, and socio-cultural distributions, diffusions, and change. In this first chapter, you can take a short, self-administered exam to test your knowledge of mapping and map reading. After you have completed the short exam in the Preface, you are ready to take on the Mapping Exercises in the main chapters of the Study Guide. As a suggestion, you might consider removing each of the maps and photocopying them so that you have plenty of copies for studying later.

In addition, this preface, like all the chapters that follow, is designed to be used in conjunction with the Marston, Knox, Liverman, Del Casino, and Robbins *World Regions in Global Context: Peoples, Places, and Environment* 4/e (2011). Throughout the guide, you will be referred to the main textbook for the data to be used in making your maps and for discussions related to the types of maps geographers use. The following introduction is broken down into two sections—understanding location and testing your map knowledge. The first section contains a number of key map concepts that will be valuable as you continue working through this workbook. This first section complements the Appendix in the text, which is titled "Maps and Geographic Information Systems."

Understanding Location[1]

Often, location is *nominal,* or expressed solely in terms of the names given to regions and places. We speak, for example, of Washington, D.C., or of Georgetown, a location within Washington, D.C. Location can also be used as an *absolute* concept, whereby locations are fixed mathematically through coordinates of latitude and longitude. **Latitude** refers to the angular distance of a point on Earth's surface, measured in degrees, minutes, and seconds north or south from the equator, which is assigned a value of 0º. Lines of latitude around the globe run parallel to the equator, which is why they are sometimes referred to as *parallels.* **Longitude** refers to the angular distance of a point on Earth's surface, measured in degrees, minutes, and seconds east or west from the *prime meridian* (the line that passes through both poles and through Greenwich, England, and which is assigned a value of 0º). Lines of longitude, called *meridians,* always run from the North Pole (latitude 90º north) to the South Pole (latitude 90º south). Georgetown's coordinates are 38º55' N, 77º00' W (**see Marston et al. 2011, p. A-3, Figure A.3**).

Thanks to the Global Positioning System (GPS), it is very easy to determine the latitude and longitude of any given point. The **Global Positioning System** consists of 21 satellites (plus 3 spares) that orbit Earth on precisely predictable paths, broadcasting highly accurate time and locational information. The U.S. government owns the GPS, but the information transmitted by the satellites is freely available to everyone around the world. All that is needed is a GPS receiver. Basic receivers cost less than $100 and can relay latitude, longitude, and height to within 10 meters day or night, in all weather conditions, in any part of the world. The most precise GPS receivers, costing thousands of dollars, are accurate to within a centimeter. GPS has drastically increased the accuracy and efficiency of collecting spatial data. In combination with **geographic information systems** (GIS) and **remote sensing** (the collection of information about parts of Earth's surface by means of aerial photography or satellite imagery), GPS has revolutionized mapmaking and spatial analysis (see, for example, **Marston et al., 2011, p. 154, Fig. 5.2,** along with the satellite imagery presented at the beginning of all the chapters).

Location can also be *relative,* fixed in terms of site or situation. **Site** refers to the physical attributes of a location: its terrain, its soil, vegetation, and water sources, for example. **Situation** refers to the location of a place relative to other places and human activities: its accessibility to routeways, for example, or its nearness to population centers. Washington, D.C., has a low-lying, riverbank site and is situated at the head of navigation of the Potomac River, on the eastern seaboard of the United States.

Finally, location also has a *cognitive* dimension, in that people have cognitive images of places and regions, compiled from their own knowledge, experience, and impressions. **Cognitive images** (sometimes

---

[1]This section and the following section, "Understanding Maps," were originally part of Marston, Knox, and Liverman (2002) *World Regions in Global Context: Peoples, Places, and Regions, 1st Edition,* p. 5–11. They have been modified with examples relevant from the 4th Edition.

referred to as *mental maps*) are psychological representations of locations that are made up from people's individual ideas and impressions of these locations. These representations can be based on people's direct experiences, on written or visual representations of actual locations, on hearsay, on people's imagination, or on a combination of these sources. Location in these cognitive images is fluid, depending on people's changing information and perceptions of the principal landmarks in their environment. Some things, indeed, may not be located in a person's cognitive image at all!

Now turn to the Appendix that begins on page A-1 (Appendix) in the back of the Marston textbook and read the entire section. When you complete that section, you are ready to answer the questions that complete this particular chapter. Enjoy and good luck.

Testing Your Map Knowledge

Fill in the blanks and answer the following questions based on your reading.

1. Lines of ______________ run east to west and measure north to south, while lines of ______________ run north to south and measure east to west.
2. 0^0 longitude is called the __________________, and 0^0 latitude is called the _____________.
3. When locating a place solely based on its name, we refer to that as its _____________ location. On the other hand, when a place is given by its coordinates in latitudinal and longitudinal space, we refer to that as its ____________ location.
4. A _________________________ is an effective tool for measuring the exact location of particular place through the use of satellite technology.
5. The relative location of a place is fixed in terms of its __________ and ____________.
6. Site refers to _____________________________, while situation is used to refer to _________________________.
7. Our psychological representations of the world are often called _______________, which are also called mental maps.
8. The Peter's projection is a(n) _______________ projection that distorts _____________.
9. ____________________ is a term that refers to the collection of information about parts of the earth's surface by means of aerial photography or satellite imagery.
10. Which of the following is the smallest scale?
    (a) 1/10,000 (b) 1/100,000 (c) 1/1,000,000 (d) 1/1,000
11. Using Map A.8, answer the following questions.
    (i). Which of following countries has a natural population increase above 3?
    (a) Iran (b) China (c) India (d) Japan
    (ii). Calling someone in which of the following cities costs the most from the United States?
    (a) Stockholm, Sweden (b) Sydney, Australia
    (c) Tokyo, Japan (d) Seoul, South Korea
12. The most time-consuming aspect of GIS work is called _________________, which is when the geographer enters data into the system.
13. __________________ utilizes census and commercial data to conduct studies, such as market research.

CHAPTER 1

World Regions in Global Context

Learning Objectives

After reading and studying this chapter, you should be able to:

1. comprehend the meaning of a global geography framework and the concepts geographers use to explain interdependence between world regions and between global processes and particular places;
2. compare and contrast the different ways in which geographers have examined the earth as a system and the relationship between humans and the environment;
3. distinguish the differing processes that have led to the current physical formation of the earth;
4. comprehend the difference between weather and climate and how these two processes impact the organization of various ecosystems;
5. understand the current political and economic configuration of the world-system in a historical perspective;
6. have a working knowledge of the processes of globalization and the impact of those processes on place;
7. evaluate the changing geography of world regions in relation to the complex changes found in the world system over time; and
8. have a set of tools for evaluating the future of the global connections and relations in a world regional geography context.

Vocabulary Review

1. Geographers often use the phrase ____________________ to designate a large-sized territory made up of a number of different ____________________ that have similar attributes.
2. The Cold War refers to a period of time when ____
____.
3. When groups of people exhibit similar attributes, such as religious identities, within state boundaries we can refer to this as the process of ____
____.
4. The term leadership cycle refers to ____________________.
5. A farmer in Tanzania is working in the ____________________ sector of the economy, while a woman working in a factory making sneakers in Indonesia is working in the____________________ sector of the economy.
6. The ____________________ is a term that refers to the process by which high birth and death rates are replaced by low birth and death rates.
7. ____________________ refers to a country's total value of all materials, foodstuffs, goods, and services produced by a country. ____________________, on the other hand, includes all of the above as well as the value of income from abroad.
8. When one country asserts that a minority population living outside its formal borders belongs to it, that is called ____________________.
9. A ____________________ is networks of labor and production processes that originate in extraction or production of raw materials and end in the consumption of finished products.

10. The term ____________________ refers to those countries that are predominantly located in the southern hemisphere. This region is also sometimes referred to as the majority world because it dominates global population.

11. ____________________ refers to the theory of how Earth's outer layer has been structured over time.

12. ____________________ refers to daily atmospheric conditions, while ____________________ refers to the average of atmospheric conditions over an extended period of time.

13. Where air rises vertically due to intense solar heating near the equator we see the development of the ____, often abbreviated as ____________________.

14. The name assigned to the classification of ecosystems into major types of vegetation and climate is a ____________________. A person interested in studying the different biomes and/or ecosystems would probably study ____________________.

15. ____________________ is a term often used to describe the process by which increasing levels of carbon dioxide and other gases are trapping heat in the earth's atmosphere. This process is sometimes called the ____________________.

16. In contrast to an ecosystem, a ____________________ is defined as the increasing political and economic interdependence between countries at a global scale.

17. Centers of socio-cultural and political-economic production in the pre-Columbian Americas, for example, are historically known as ____________________.

18. World empires are important because they facilitated the large-scale distribution of ideas and goods, often through the twin processes of ____________________ and ____________________.

19. When a previously homogenized group becomes spatially dispersed, we often refer to that group as a ____________________. On rare occasions one homogeneous group will occupy the boundary of a nation-state, but in most cases states are multinational, and ____________________, the feeling of belonging to a nation, must be actively constructed.

20. ____________________ is the term one would use to describe the domination of one set of peoples over the world economy.

21. A hegemony will often impose a new ____________________ on its colonial holdings through the development of geographically specific patterns of specialized production.

22. Even though most countries today have been decolonized, one can still find many instances of what is now referred to as ____________________, through which powerful states in the core still maintain influence over the economy and politics of peripheral countries.

23. The term BCE stands for ____________________. The __________ period in World History began in 500 BCE and ended in 500 CE.

24. Global neoliberal economic policies are often carried out by two macro-level international organizations called the ____________________ and the ____________________, which is more commonly known by its acronym ____________________.

25. ____________________ is supportive of decreases in government spending as well as the privatization of all government holdings.

26. A ____________________ or TNC is an organization that has investments that transect international borders and involve the labor and resources of numerous countries.

Activities & Problems

1. According to Figure 1.21, which world regions have more than 25% of their population currently using the Internet?

____

2. Based on your answers to question 1, provide a few hypotheses as to why most of Africa remains without Internet connections.

____

____

____

3. Referring to Figure 1.8 in your text identify the types of boundary between the following plates.
   (a) North American Plate and the Juan de Fuca Plate ____________________
   (b) South American Plate and the Nazca Plate ____________________
   (c) Philippine Plate and the Eurasian Plate ____________________
   (d) Indo-Austrian Plate and the Eurasian Plate ____________________
   (e) The Pacific Plate and the Philippine Plate ____________________

4. Identify the following features on the figure below: (a) tropic of Capricorn; (b) tropic of Cancer; (c) Hadley cells; (d) ITCZ; (e) trade winds; (f) westerlies; (g) southeast trade winds; and (h) north-east trade winds.

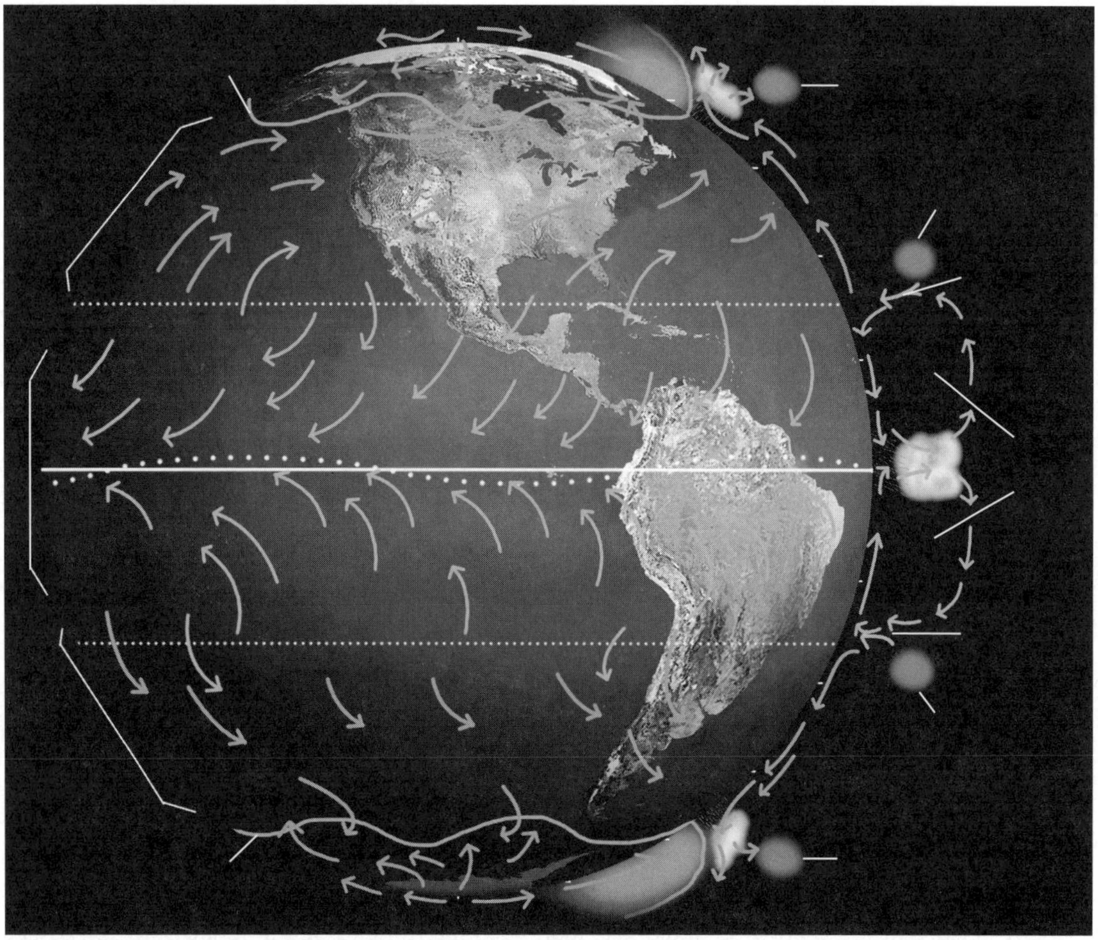

5. Based on the figure in question 4. explain why the ITCZ is located near the equator.

____

____

____

6. Based on Figure 1.5 in your textbook name three central African countries that have a tropical wet with no dry season climate: ____________________, ____________________, and ____________________.

7. Using the same figure as in question 4, name three countries in Southeast Asia that have tropical savanna climatic regions: ____________________, ____________________, and ________________.

8. Based on Figure 1.16 in your textbook compare the spatial extent of France in 1714 and in 1914. Do the same for Spain. What is the major difference between the two?

____

____

____

____

____

____

Review Exam

1. What is meant by a global geography framework?

____

____

____

____

____

2. What defines the differences between the paleolithic and the neolithic periods in world history?

____

____

____

____

____

3. Briefly discuss how the process of globalization has fostered interdependence between places. Use an example to support your argument.

____

____

____

4. According to the World Commission on the Environment and Development, what is sustainable development?

5. What is the difference between Gross National Product (GNP) and Gross Domestic Product (GDP)?

6. What accounts for the demographic transition?

7. What happens when a country gets caught in a "demographic trap"?

8. What is meant by the term "sense of place"?

9. What does a regional geographer study? How does that differ from the study of human or physical geography?

10. What is meant by the term "spatial justice"? How might a geographer study spatial justice in relation to gender relations globally? Use examples to support your argument.

11. Briefly describe the theory of continental drift.

12. What is meant by the term biodiversity? Why do we have such diverse bioregions today?

13. Using Figure 1.40, discuss some examples of global patterns of invasive species and briefly discuss their impact.

14. Based on Figure 1 in the box "Six Key Factors of Globalization," what can you say about world trade interconnections? What trends do you see? What regional disparities are apparent?

15. What does the future hold for global population geography in 2025 and 2050?

16. Compare and contrast the terms "nation" and "nation–state." What role does "nationalism" play in the formation of these particular entities?

17. Describe the geographies of the Cold War.

18. What is meant by the term "international division of labor" in the context of colonization?

19. Briefly compare and contrast the perspectives on the world in 1491 from at least two different places.

20. What impact did colonization have on the transfer of crops around the world? Using the potato as an example, briefly describe the historical geography of that crop's movement across the globe.

21. Why are experts predicting that energy consumption in the global south will double by the year 2025?

22. Briefly describe the spatial distribution of climatic zones at the global scale in relation to the major world regions covered in the text.

____

____

____

____

23. Describe the relationships between climatic zones and world ecosystems at a global scale.

____

____

____

____

____

24. The generic term for the process by which one ethnic group seeks to create autonomy for themselves within a national context is often called ____________________.

(a) irredentism (b) regionalism (c) sectionalism (d) nationalism

25. When one state exerts political and legal domination over another territory, this is called:

(a) economies of scale (b) imperialism
(c) jingoism (d) colonialism

26. Which of the following terms is used to describe a region that seeks a high degree of autonomy from that national state

(a) cultural fault line (b) racialization
(c) regionalism (d) supranationalism

27. When a collection of national states formally come together to address a common goal, we call the resultant structure a ____________________.

(a) world region (b) supranational organization (c) global community

28. Which of the following DOES NOT reflect a core concept of a global geography framework ____________________?

(a) global interdependence (b) systemic relations (c) local isolation (d) ongoing change

29. When one country asserts a historic and cultural right over a minority living outside of its current national borders, it is making a(n) ____________________ claim.

(a) sovereignty (b) colonialist (c) imperialist (d) irredentist

30. Someone working in the area of knowledge and information technology is working in which sector of the economy?

(a) primary (b) secondary (c) tertiary (d) quaternary

31. By the year 2030, it is estimated that global energy consumption will surpass how many quadrillion BTUs?

(a) 280 (b) 420 (c) 510 (d) 720

32. ____________________ is the process by which trade winds pick up water over the ocean and then drop that water as rain at the top of coastal mountains.

(a) biogeography
(b) orographic lifting
(c) biome development
(d) global warming

33. Which of the following transnational organizations provides loans to governments throughout the world?

(a) IMF
(b) TNC
(c) World Bank
(d) GATT

34. ____________________is a term that is used by geographers when developing a classification scheme for places or areal units.

(a) globalization
(b) regionalization
(c) homogenization
(d) hyperglobalization

35. The term "diaspora" refers to which of the following?

(a) the unification of a formally homogeneous group
(b) the unification of formally heterogeneous groups into a unified group
(c) the spatial dispersion of a previously homogeneous group
(d) the spatial dispersion of a number of heterogeneous groups

36. During the ____________________ incoming solar energy is greatest in the Northern Hemisphere?

(a) vernal equinox
(b) winter solstice
(c) autumnal equinox
(d) summer solstice

37. Which of the following is NOT a key factor in the process of "globalization"?

(a) a new international division of labor
(b) the internationalization of finance
(c) transnational corporations
(d) mercantalism

Mapping Global Life Expectancy[1]

1. In this exercise, you are going to examine average life expectancy globally. Begin by taking a clean copy of the map provided and drawing boundaries around the 18 world regions listed in the "2009 World Population Data Sheet" (WPDS) located at the back of this guide. (NOTE: Aside from "North America" and "Oceania" the other 16 regions are listed as subregions of larger continental spaces, such as Africa, Europe, Latin America and the Caribbean, etc.)
2. Now, looking at the WPDS find the column labeled "Life Expectancy at Birth (years), Both Sexes." Using a table, list the world regions and their corresponding life expectancy figure from greatest to least. Rank these expectancies from greatest to least, placing a number 1 next to the greatest life expectancy and an 18 next to the lowest.
3. On your map of world regions, label each region with its corresponding number (and life expectancy mean in brackets). Answer the following questions:

   (a) Which world regions have the lowest life expectancy? Which regions have the highest?

   ____

   ____

[1]Answers to Map Exercises will vary and therefore no answers are given in the corresponding Answer Section below.

(b) Do you notice a spatial pattern in the statistics? Are there any distinctive lines that can be drawn, globally, between the top five regions and the five lowest regions? If so, what are they?

(c) Going back to your original statistics, which specific countries seem to have the lowest life expectancy in relation to their regions and to the world (you can calculate a world mean using the averages of all the world regions)? Why do you think these countries stand out? Explain.

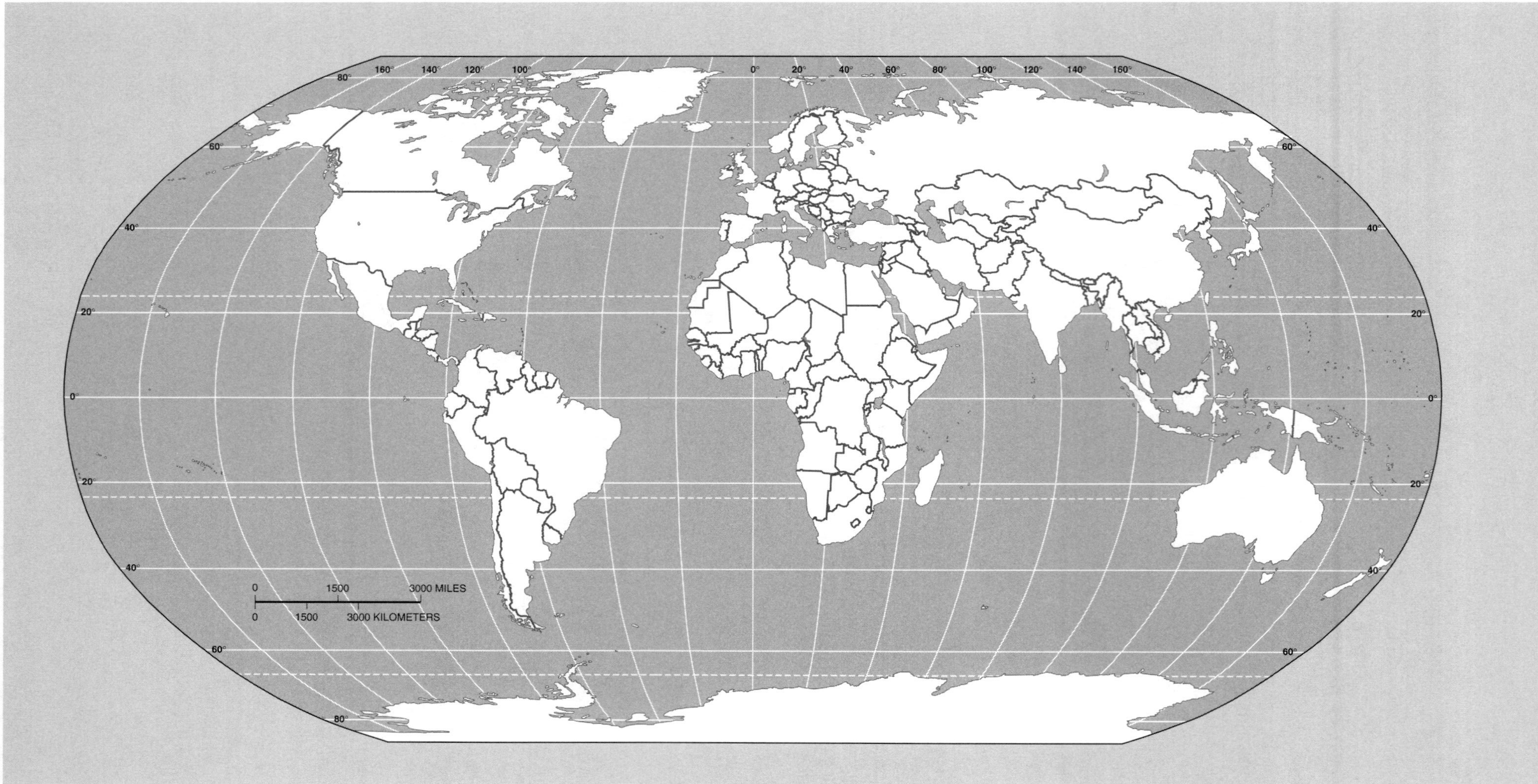

MAP 1.1

CHAPTER 2

Europe

Learning Objectives

After reading and studying this chapter, you should be able to:

1. understand the complex physical geography of Europe as well as the important relationships between people and their environments;
2. have a working knowledge of the relationship between the rise of Europe as a major world region and the development of capitalism;
3. distinguish the distinct cultural patterns that are part and parcel of the complex population geography of Europe;
4. compare and contrast the development policies of "western" and "eastern" Europe;
5. examine the distinct patterns of population and migration that distinguish this region;
6. comprehend the dynamics of regional integration and difference in Europe; and,
7. understand the importance of political and economic integration in the region and the pace at which that integration is occurring.

Vocabulary Review

1. The four main physiogeographic regions of Europe are ____________________, ____________________, ____________________, and ____________________.
2. Where seas have risen to meet glaciated landscapes one can often find ____________________.
3. In the Netherlands the local population perfected the practice of draining lands through the use of windmill pumps, drainage ditches, and canals. The results of their efforts were flat, fertile, stone-free land, or what is also called a ____________________ landscape.
4. Industrialization in Europe had profound impacts, including the production of ____________________, which resulted from heavy doses of atmospheric pollution related to factory production.
5. Eastern Europe was often considered to be a ____________________ between the Soviet Union and western Europe during the Cold War period. The countries caught between the two powerful blocks were often thought of as ____________________ states of the Soviet Union.
6. Despite a socialist or communist rhetoric, the states of eastern Europe and the Soviet Union developed into ____________________ economies, which were highly bureaucratic and state centered.
7. The process by which larger states are broken up into smaller states, particularly around ethnic groups, is called ____________________.
8. In some cases ethnic minorities do not break away from a country but instead remain in ____________________ within another state.
9. In some cases extreme nationalism, which is sometimes referred to as chauvinism, has led to the extreme practice of ____________________, whereby one ethnic group is forcibly evicted from a particular place.
10. Immigration creates both positive and negative results. In some cases, immigration of ethnic minorities into countries in Europe has led to a rise in ____________________, hate or fear of a minority group.

11. During the early part of the 20th century, industrial employment was scaled back in places such as northern England, south Wales, and central Scotland. This process is called ____________________, and it often is accompanied by ____________________, which includes the selling off of assets such as factories or equipment.

12. The ____________________ is the modern core region of Europe that stretches from London to Paris to Berlin.

13. The development of a centralized, autocratic government, which values nation and race over individuals, is often defined by the term ____________________.

14. In this region, the emergence of the ____________________ has supported the growth of services and benefits in the areas of high quality medical care, public transportation systems, and social housing.

15. ____________________ is defined as a way of life based on the herding and grazing of animals. More often than not regions where such practices are present have relatively low-density populations.

16. As one moves north in Europe, the vegetation patterns become consumed by gray lichens and dwarf willows and birches, and forests cover over half the land. This area is best representative of a ____________________ biome.

17. On the Danubian Plain, particularly in the Hortobágy area, one would find semiarid, treeless grassland, also representative of a ____________________ biome.

18. In the Mediterranean region of Europe successive empires carved out large estates, known as ____________________, where peasants were often forced to work. Land that was not part of these larger estates was sometimes parceled off into smaller properties by peasant farmers who called them ____________________.

19. Intermediary cities, or ____________________, were seaport towns that served as trading and trans-shipment points throughout Europe.

Activities & Problems

1. On the following key identify the four major physiographic regions by name: (a) northwestern uplands; (b) north European lowlands; (c) central plateaus; and (d) Alpine system

2. Between 1945 and 1972 a significant amount of the migrant population came from outside Europe. According to your reading of Figure 2.28 in your text, what non-European population was the largest migrant group into Germany? ___________________ Into France?

____

3. Between 1945 and 1972, there was a significant amount of labor migration between countries in Europe (see Figure 2.28 in your text). Internal migration within the region, however, was not even, and some countries had much higher numbers of out-migrants. Name the seven countries within Europe that witnessed significant out-migration in this particular period: ___________________,
___________________, ___________________, ___________________,
___________________, ___________________, and ___________________.

4. Give two reasons why the internal migration patterns of Europe between 1945 and 1972 look the way they do.

5. Explain why the population pyramid of Germany (Figure 2.27 in your text) has the shape it does.

Review Exam

1. Explain the relationship between trade and the "Age of Discovery."

2. What is a command economy?

3. What impact has the European Union had on agriculture in Europe?

4. What four advantages have allowed Europe's "Golden Triangle" to thrive?

5. Describe the characteristics of the "Southern Crescent." Where is it and why has it become a place of relative prosperity?

6. Briefly describe the diversity of Europe's Muslim population, considering why there has been an increase in Muslim in-migration to Europe since World War II.

7. Which language group is most commonly used on the Iberian Peninsula?

(a) Romance (b) Germanic (c) Celtic (d) Baltic

8. A person living in Sicily would be living in which physiogeographic region?

(a) northwestern upland

(b) north European lowland

(c) central plateaus

(d) Alpine system

9. According to Figure 2.16, which of the following regions are eligible for aid from the European Union's regional development program?

(a) southern portion of the United Kingdom

(b) southern portion of Spain

(c) Switzerland

(d) most of Western Germany

10. Which of the following areas in Europe still had not been impacted by industrialization by 1914?

(a) northern Ireland

(b) western France

(c) northern Italy

(d) southern Italy

11. Portuguese workers have tended to seek employment in which country?

(a) Spain (b) United Kingdom (c) France (d) Germany

12. The ____________________ movement was based on the conviction that all of nature, as well as human beings and societies, could be understood as a rational system.

(a) poststructuralism (b) enlightenment (c) communism (d) postmodernism

13. Which of the following is not true about agricultural production in the Golden Triangle of Europe?

(a) it is highly intensive

(b) it is geared toward the rural hinterlands

(c) it is focused on fresh dairy, vegetables, and flowers

(d) it is used to supply the city

14. The growth of the Southern Crescent, a secondary core region in Europe, has resulted from which of the following?

(a) the decentralization of industry from northwest Europe

(b) the influx of capital from Middle Eastern oil companies

(c) the expensive labor in the southern part of the region

(d) the slow rate of urbanization in the region

15. Which of the following cities is part of the emerging axis of growth in Europe?

(a) Dublin

(b) Glasgow

(c) Hamburg

(d) Madrid

16. Which of the following has hindered the reintegration of eastern Europe into the overall regional economy?

    (a) ethnic conflict in Bosnia-Herzegovina, Croatia, and Yugoslavia

    (b) low levels of education in the Baltic states

    (c) agriculture is now geared toward serving the international market

    (d) an increase in wage and price controls in eastern Europe

For each of the following statements, write T for "True" or F for "False" on the space provided.

17. As part of the Common Agricultural Policy (CAP), the European Union supports wholesale prices of agriculture. ___________________

18. Historically, the staple economy of the Alpine region has been agriculture. ___________________

19. The Marshall Plan was designed to promote command economies in western Europe. ___________________

20. A citizen of Greece would most likely be speaking a Romance-based language. ___________________

21. Road building represents over 50% of the European Union's priority project budget. ___________________

22. Xenophobia is non-existent in Europe today. ___________________

23. The Mediterranean climate tends to be cool in the winter. ___________________

24. Europe is home to approximately 15 to 20 million Muslims today. ___________________

25. Heathlands consist of open land with coarse soil and poor drainage. ___________________

Mapping Key Places

Identify the following features on Map 2.1 (countries and cities) and Map 2.2 (landforms and waterways). On your political map, Map 2.1, label the major cities by writing the city names next to their respective symbols. When mapping some of the landforms, you may need to create some sort of symbol system that portrays the landform, such as a ^^^^^^^^^^^^ for a mountain range.

Countries	**Major Cities**	**Landforms**	**Waterways**
As listed in Chapter 2 of your textbook (Figure 2.1)	• Amsterdam • Athens • Barcelona • Berlin • Copenhagen • London • Madrid • Milan • Munich • Paris • Prague • Rome • Venice • Vienna • Warsaw • Zurich	• Alps Mountains • Appenines Mountains • Carpathian Mountains • Pyrennees Mountains • Balkan Peninsula • Greek Peninsula • Iberian Peninsula • Italian Peninsula • Scandanavian Peninsula • Island of Sicily • Island of Sardinia • Island of Crete	• Danube River • Dnieper River • Elbe River • Rhine River • Seine River • Thames River • Adriatic Sea • Baltic Sea • Black Sea • Mediterranean Sea • North Sea • Strait of Gibraltar • Bosporus and Dardenelles Straits • North Sea • English Channel

Map 2.1 Countries and Cities

Map 2.2 Landforms and Waterways

Mapping Spatial Distribution of GNI PPP/Capita

Goal: In this exercise you will create a choropleth map of GNP/capita for the countries of Europe. A choropleth map is a map that uses shading to illustrate difference and proportion or magnitude in values in a given region. An excellent example of a choropleth map is Figure 1.22 on page 25 of your Marston *et al.* (2011) textbook.

1. Open this guide to the WPDS at the back of this text.
2. Find the countries of Europe and scroll across to the column titled GNI PPP (US$) per capita 2008.
3. On a separate piece of paper order the countries by GNI PPP per capita from least to greatest.
4. A choropleth map is based on categories of data. Looking at your data as they are listed from least to greatest, organize the data into five distinct categories.
5. Next, create a key for each one of the data categories and shade that category accordingly (i.e., the lightest shade should be used for the smallest category). The key should also be placed on your map (see discussion above and samples in the text of choropleth mapping).
6. Finally, create a map of your data by labeling the countries according to your categorization using a copy of Map 2.3, the Political Map of Europe, which is provided for you on the next page. Place a key with your categories somewhere to the side of the map. Also, give your map a title.
7. Describe the pattern that you see.

8. Offer several hypotheses that might explain the pattern (cite key points in the textbook that help support your hypotheses).

40° E
30° E
20° E
10° E
0°
10° W
20° W
30° W
Arctic Circle
60° N
50° N
40° N
N
0 200 400 Miles
0 200 400 Kilometers

Map 2.3

CHAPTER 3

The Russian Federation, Central Asia, and the Transcaucasus

Learning Objectives

After reading and studying this chapter, you should be able to:

1. understand the transitionary nature of this region's wealth and resources;
2. appreciate the complex human-environment relations in this region and the relationship between those relations and environmental problems today;
3. comprehend the history of the Soviet State and the relationship between Russia and the former Soviet Republics;
4. understand post-Soviet economic development trajectories;
5. comprehend the historical changes in the population geography of the region and their impact on the current geopolitics of conflict and tension throughout the region; and,
6. understand the complex future trajectories this region might take.

Vocabulary Review

1. In the ____________________ region it is unlikely that there would be any agricultural production or forestry.
2. The largest physiogeographic zone in the region is the ____________________, which refers to an area that is made up of large tracts of spruce, fir, and pine trees. The ____________________ region is a flatter landscape devoid of trees and dominated by tall grasses.
3. Black earth, known by its technical name ____________________, can be found throughout the ____________________ region and has a high level of fertility.
4. In contrast to the Czarist ____________________ state, which was highly centralized, the Leninist model for the Union of Soviet Socialist Republics (USSR) was a ____________________ state in which power was to be devolved to the local republic level.
5. A ______________________ is founded on the ideas of an open civil society where diverse groups can actively participate in political life.
6. In the post-Soviet era, despite attempts to liberalize the state, ____________________ organizations, such as nongovernmental organizations, business organizations, and pressure groups, have not emerged.
7. In the northernmost part of this world region the ground is constantly frozen, a condition known as ____________________.
8. Under ____________________ basic services, such as housing, health care, and education, were provided to the citizens of the Soviet Union at little to no cost.
9. During the Soviet period large tracts of the eastern steppe region were cultivated and wheat was grown in large quantities. Lack of rainfall in the region, however, demanded that farmers turn to ____________________ techniques and irrigation, which included the constant weeding of fields and leaving the stubble of plants to trap snow in the winter.
10. Uzbekistan has suffered from large-scale agricultural production of cotton. In particular, ____________________, which is caused when water evaporates and leaves salt behind on top of the soil, has slowly diminished the productive capabilities of cotton farmers in this now-independent state.

11. Because the Soviet industrial economy was not based on free markets, the government organized ____________________, a set of regionally based production facilities to maximize productivity.

12. Central Asia was tied into the world economy through the 11th century because the ____________________ ran directly through, and between, the small agricultural oasis communities of this subregion.

13. In the post-Soviet period this region has been marked by several ____________________ and ____________________ claims, the latter of which includes Chechnya's attempt for political independence and the former of which includes the claims of Tajikistan and Kazakhstan over territories currently outside their own political borders.

14. The Iron Curtain was a metaphor used to describe the territorial extent of Soviet control in ____________________ after 1948 and prior to the breakup of the Soviet Union.

15. The ____________________ form a physiogeographic boundary between the Russian and West Siberian plains.

16. The ____________________ is an inland sea that is choked off by the Bosporus and Dardanelles Straits as well as the Sea of Marmara.

17. A small group of individuals who have extreme power are sometimes referred to as ____________________.

18. The ____________________, or state bureaucrats, were key actors in the organization and development of the Soviet State.

19. In an attempt to combat the economic upheaval caused by the breakup of the Soviet Union, the Russian Federation led a coalition of new states in the formation of the ____________________, or CIS, which was a forum to discuss economic and political problems in the region.

20. The ____________________ is a region made up of former Soviet Republics with large numbers of ethnic Russians in which the Russian Federation still wants to assert its ongoing influence.

Activities & Problems

1. Using Figure 3.19 in your text, answer the following questions:
   (a) In what century did the Russian State expand out to the Caspian Sea? ____________________
   (b) In what century did the Russian State expand out to the Black Sea? ____________________
   (c) In what century did the Russian State expand out of the Sea of Okhotsk? ____________________
   (d) In what century did the Russian State acquire the land to establish the city of St. Petersburg? ____________________

2. Using Figure 3.20 in your text, answer the following questions:
   (a) What three territories were incorporated into the USSR in 1940? ____________________, ____________________, ____________________
   (b) What three eastern European countries were incorporated into the Soviet block in 1947? ____________________, ____________________, ____________________
   (c) When did the Soviet Union occupy the eastern part of Austria? ____________________
   (d) During what time period was Albania part of the Soviet bloc? ____________________
   (e) What three countries lost part of their territory to the Soviet Union between 1940 and 1947? ____________________, ____________________, ____________________

3. Using Figure 3.17, answer the following:
   (a) Which river bordering the People's Republic of China is heavily contaminated?____________
   (b) Which city located in the northern reaches of the country bordering the Barents Sea is located within a region of catastrophic environmental degradation? ____________
   (c) Nuclear weapons have been tested on a northern island in Russia, which is located between what two bodies of water? ____________, ____________
   (d) Name at least three bodies of water where the Russians have stored nuclear waste?
   ____________, ____________, ____________

4. Compare and contrast the different vegetation patterns found in tundra, taiga, and steppe landscapes of this region.

5. Using Figure 3.29 discuss the geographies of seccessionist and irrendist claims in the Post-Soviet Union.

Review Exam

1. Describe the physiographic regions of the Russian Federation, Central Asia, and the Transcaucaus.

2. Under Stalin how did Soviet industrial policy change? What were the impacts of that change?

3. Why is it argued that the future of this region may take a number of different paths?

4. Briefly describe the ongoing conflict in Chechnya.

5. What are some of the causes of crime and corruption in Russia during the period of market reform?

____

____

____

____

____

____

____

6. Which of the following is true about the Central Region?
   (a) It is highly urbanized.
   (b) It is highly rural.
   (c) Moscow is the third largest city in the region.
   (d) About 75% of the region is forested.
7. Which of the following is not true about post-Soviet Moscow?
   (a) The transition has been smooth with little conflict.
   (b) The housing market has slowly risen to meet the demand of the growing population.
   (c) The construction of new homes has grown uncontrolled in the forest protection belt that surrounds the city.
   (d) There has been little appearance of Western-style stores or shops.
8. An area of sparse vegetation with no agricultural or forestry production capabilities is called a ____________________.
   (a) steppe (b) chernozem (c) taiga (d) tundra
9. Which region is not considered to be "inside" the old Soviet Union's manufacturing belt?
   (a) Eastern Ukraine (b) Volga (c) Far East (d) Moscow
10. In 1994, Russian troops invaded which of the following republics, kicking off seven years of warfare?
   (a) Georgia (b) Chechenya (c) Dagestan (d) Adygeya
11. Which of the following was one of the major reasons for the early expansion of the Russian Empire out of Muscovy and into Siberia?
   (a) desire for oil reserves
   (b) desire for an outlet to the Pacific Ocean
   (c) desire to obtain more forest resources, such as furs
   (d) desire to trade more effectively with the Chinese

12. Which of the following regions is not considered an ethnonational hotspot in the Transcaucasus region?
    (a) Ukraine
    (b) Azerbaijan
    (c) South Ossetia
    (d) Tajikistan
13. What is the zone of boreal coniferous forest called?
    (a) steppe (b) taiga (c) tundra (d) chernozem
14. Which of the following is true of the Aral Sea?
    (a) It has shrunk by more than 90% of its surface area.
    (b) It has increased by more than 40% of its surface area.
    (c) Its sea level has dropped by more than 33 meters.
    (d) Its sea level has risen by more than 33 meters.
15. Which of the following endangered waterscapes is known as the world's deepest lake?
    (a) Lake Baykal (b) The Aral Sea (c) The Black Sea (d) Lake Tiaga

For each of the following statements, write T for "True" or F for "False" on the space provided. Questions 16-20 are based on Figure 3.28, while questions 21-25 are based on Figure 3.29.

16. Armenian speakers are located primarily in the northeast of Russia. ____________________
17. Most of the peoples of the Central Region of Russia speak Slavonic-based languages. ____________________
18. Altaic languages are spoken along the coast of the Caspian Sea and east toward South Asia. ____________________
19. Caucasian speakers can be found throughout the northernmost reaches of Russia. ____________________
20. Georgian is a Slavonic language, while Russian is an Altaic language. ____________________
21. North Ossetia holds claims over South Ossetia, while South Ossetia also holds claims over North Ossetia. ____________________
22. Turkmenistan holds several claims over regions that are currently outside their political jurisdiction, including territories in Uzbekistan. ____________________
23. Kazakhstan currently holds several claims over territories found in the Russian Federation. ____________________
24. Armenia holds claims over parts of Azerbaijan. ____________________
25. The Russian Federation holds no claims over territories outside its current political border. ____________________

Mapping Key Places

Identify the following features on Map 3.1 (countries and cities) and Map 3.2 (landforms and waterways). On your political map, Map 3.1, label the major cities by writing the city names next to their respective symbols. When mapping some of the landforms, you may need to create some sort of symbol system that portrays the landform, such as a ^^^^^^^^^^^^ for a mountain range.

Countries	Major Cities	Landforms	Waterways
As listed in Chapter 3 of your textbook (Figure 3.1)	• Ashkhabad • Baku • Dushanbe • Grozny • Khiva • Krasnoyarsk • Minsk • Moscow • Murmansk • Rostov-on-Don • St. Petersburg • Tashkent • Tbisili • Vladivostok • Volgagard • Yerevan	• Altay Mountains • Caucasus Mountains • Pamir Mountains • Sayan Mountains • Tien Shan Mountains • Ural Mountains • Verkhoyansk Mtns. • Yablonovyy Mtns. • Kamchatka Peninsula • Ostrov Sakhalin Island	• Amur River • Dnieper River • Don River • Kolyma River • Lena River • Moskva River • Neva River • Ob' River • Volga River • Yenisey River • Lake Baykal • Lake Balkhash • Aral Sea • Barents Sea • Bering Sea • Black Sea • Baltic Sea • Caspian Sea • Sea of Azov • Sea of Okhotsk

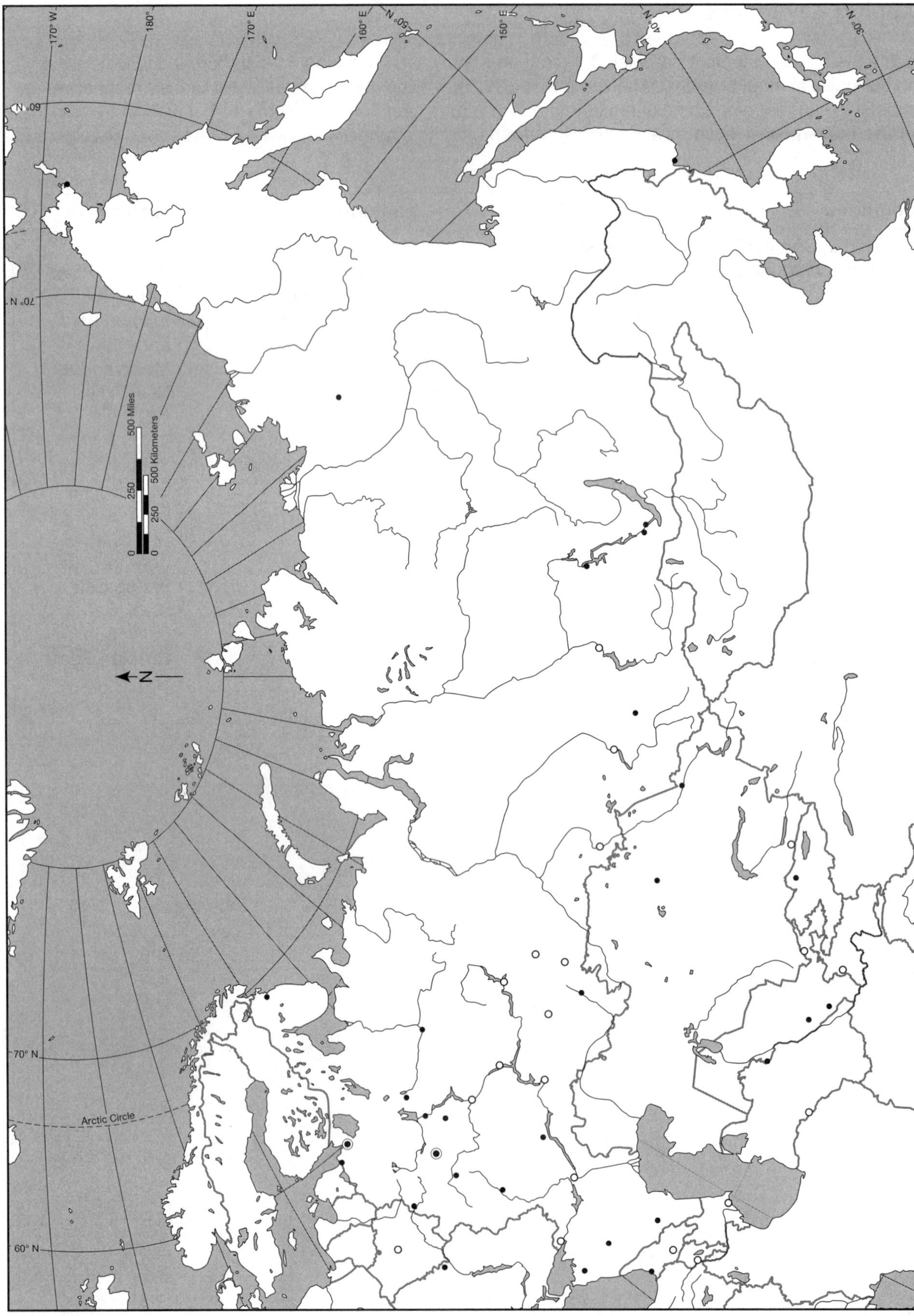

Map 3.1 Countries and Cities

Map 3.2 Landforms and Waterways

Mapping Russian Populations

Goal: Throughout the period of both the Russian State and the Soviet Union there was a significant diffusion of ethnic Russians from the Russian core into other parts of the region. In this exercise, using the blank map provided, create a choropleth map of "Russians as % of Total Population of Republic," following the directions below. These data can be found on page 104 in Table 3.1 of your textbook.

1. On a separate piece of paper order the countries by "Russians as % of Total Population of Republic" from least to greatest.
2. A choropleth map is based on categories of data. Looking at your data as it is listed from least to greatest, organize the data into three distinct categories. The only missing data is the Russian Republic, whose Russian population is equal to 81% of the total population of the republic. You should make the Russian Republic its own, fourth category, what statisticians call an "outlier," since it is so much larger than the other republics (see further discussion of choropleth mapping under "Mapping Exercise" in Chapter 2).
3. Next, create a key for each one of the data categories and shade that category accordingly (i.e., the lightest shade should be used for the smallest category). Place the key on your map somewhere to the side of the map.
4. Finally, using Map 3.3 map your data by labeling the countries according to your categorization and give the map a title.
5. Describe the pattern that you see.

____

____

____

____

____

____

____

6. Using your text and the map you have produced offer an explanation for why the map looks the way it does.

____

____

____

____

____

____

____

7. How and why might this population map change over the next 10 years? Explain.

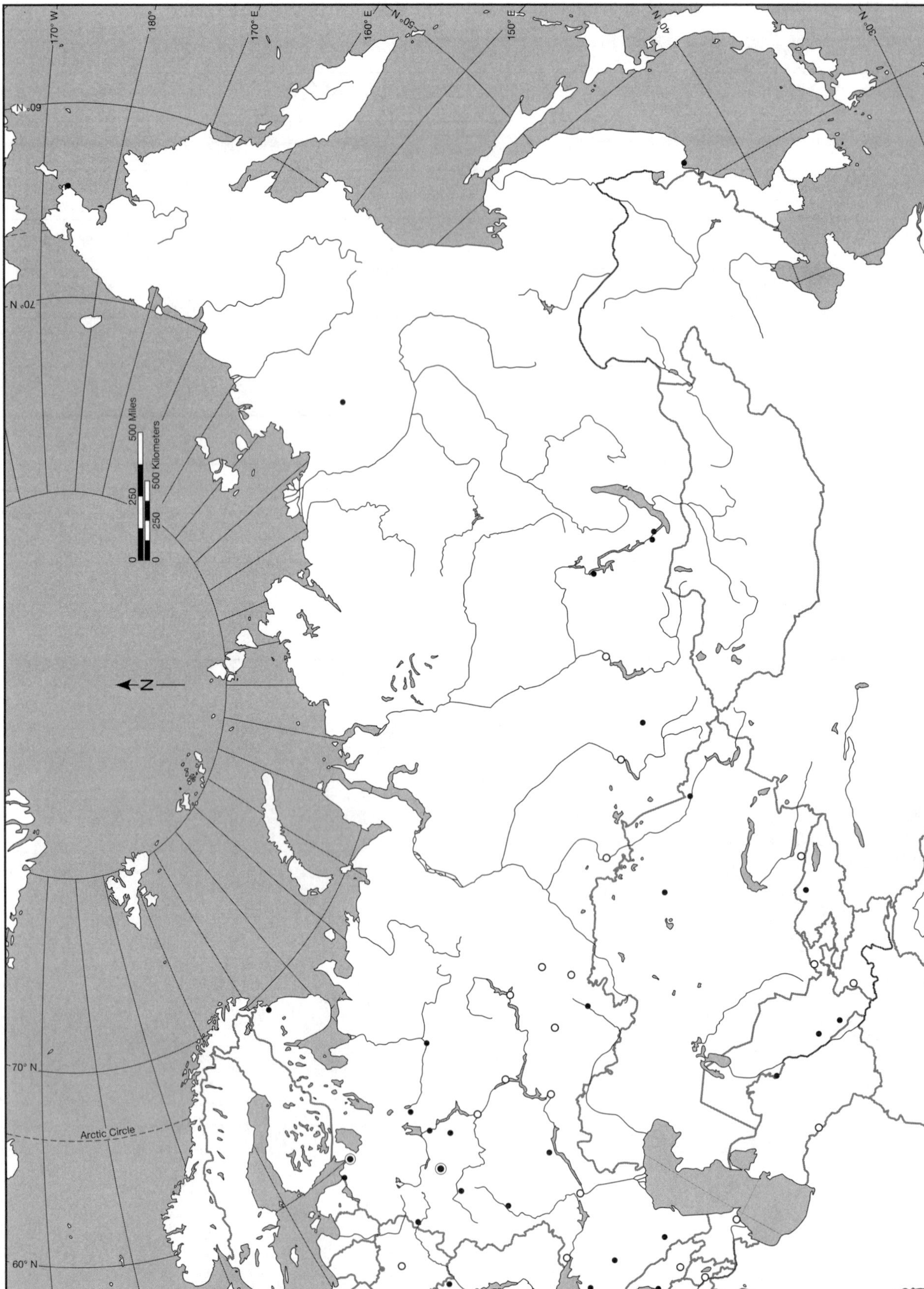
170° W
180°
170° E
160° E
50° N
150° E
40° N
30° N
60° N
70° N
500 Miles
500 Kilometers
0
250
N
70° N
Arctic Circle
60° N

Map 3.3

CHAPTER 4

Middle East and North Africa

Learning Objectives

After reading and studying this chapter, you should be able to:

1. understand this region's dynamic physical environment, despite outward appearances as a completely arid space;
2. understand the historical geography of this region and its importance as a source of domesticated plants and animals as well as world religions;
3. comprehend the basic tenets of Islam as a religious philosophy and practice and the growth of this religion in relation to Christianity and Judaism;
4. compare and contrast the differential impacts of urbanization, empire building, and European colonization on the region;
5. evaluate the complex social and spatial systems that mediate class, gender, and ethnicity in the region;
6. describe the different political tensions in the region and their concurrent political geographies; and
7. understand how the future of this region (and its potential conflicts) is tied to the importance of natural resource development—water, oil, agriculture, religion etc.

Vocabulary Review

1. The greatest defining climatic feature of the Middle Eastern and North African region is ____________________, in that the climate lacks enough moisture to support trees or woody plants.
2. A(n) ____________________ can "rise out of the desert" because of both a low water table and land erosion that exposes the water table to the surface.
3. In parts of Turkey farmers can practice "dry farming," which means they farm ____________________.
4. To combat the reduction in forest coverage, some countries are engaging in reforestation, or ____________________ programs.
5. The ____________________ were the largest indigenous ruling group in the region prior to colonization of the region by Europeans.
6. The ____________________, instituted by the British, established a homeland for Jews in ____________________ in 1917.
7. Economic activities that take place outside the "official records" of the government are called ____________________, and are often an important part of daily life for many people in the region.
8. The 1930s, which was the height of European colonization in parts of the region, was also a time when certain states, such as Turkey and Egypt, began a policy of ____________________, whereby domestic producers provided goods that were previously provided by foreigners. In the post-WWII period, on the other hand, national policies became even stricter as a number of countries followed a policy of ____________________ of all industry, turning private enterprises into state enterprises.
9. Islam is a term that means ____________________, whereas the term ____________________ refers to a person "whose duty is obedience and submission to the will of God."
10. ____________________ is often mistaken for fundamentalism, which refers to a strict adherence to the religious doctrine. But the former actually refers to ____________________.
11. The term ____________________ is often interpreted incorrectly as "holy war," while in reality the term is defined as "peaceful struggle" both internally and externally.

12. While the term ____________________ is often applied to biological relationships, it can also be extended to include ties among broader social groups, such as neighbor and friends.
13. Pastoralism is best defined as ____________________. In many cases pastoralists practice ____________________, moving herds based on seasonal changes.
14. In some cases, Muslim women wear the ____________________, a full body covering, while in other instances women wear only a ____________________ covering their head.
15. Throughout the Gulf States, which are all rich in oil reserves, a large proportion of the labor force, almost 75%, are ____________________, brought in from other countries to work in oil production.
16. The ____________________ is the sacred pilgrimage to Mecca that each financially and physically able Muslim man should make over the course of his life.
17. The Palestinians, in an act of defiance against the Israeli state, began a(n) ____________________ or uprising in the late 1980s to protest for an independent Palestine.
18. ____________________, or revenue that results from the sale of oil, are often reinvested into the larger world system of banking and investment.
19. ____________________ was a late 19th century movement that pushed for the development of a legal homeland for Jews.
20. Throughout Sudan, Turkey, and Iran there is a new and emergent group of individuals labeled as ____________________ because they have refugee status within their own country.
21. The process by which arid and semiarid lands become degraded by increasing forest and grassland loss is known as ____________________.

Activities & Problems

1. Based on Figure 4.18, name the colonial power that occupied the following countries:
   (a) Morocco ____________________
   (b) Kuwait ____________________
   (c) Libya ____________________
   (d) Egypt ____________________
   (e) Tunisia ____________________
   (f) Anglo-Egyptian Sudan ____________________
2. Based on Figure 4.34, answer the following questions about restrictions on women's movements.
   (a) In which countries do all of the following apply: women are secluded by some ethnic groups; there exists a compulsory government-mandated dress code; there have been attacks on women for immodest dress; and women need permission from a male relative to get a passport to travel abroad?

   ____

   (b) Which countries do not require a male relative to give permission for a woman to get a passport?

   ____

   (c) Which North African countries have reported attacks on women for immodest dress?

   ____

   (d) In which countries do none of the following apply: women are secluded by some ethnic groups; there exists a compulsory government-mandated dress code; there have been attacks on women for immodest dress; and women need permission from a male relative to get a passport to travel abroad?

   ____

3. Examining Figure 4.26 in your text, explain migration patterns in the region in 1990. When did these patterns begin to develop? What was the impetus for the migration patterns we see today?

____

____

____

____

____

____

4. Based on your reading of Figure 4.37, how did the map of Israel change between 1949, 1967, and 1996? How does the map of 2009 differ today compared with 1996?

____

____

____

____

____

____

5. As populations continue to increase, there is going to be more and more pressure on the lands of the Middle East and North Africa. In this question, you are going to: (a) calculate projected population densities for the countries of this region; (b) compare those calculations to the 2009 population density; and (c) examine the total percentage of change in population density over the two time periods.

   (a) Turn to the "2009 World Population Data Sheet" located at the back of this text. On a separate piece of paper make the following columns in this order (you may want to use a spreadsheet program, such as Excel, for this exercise): (a) country name, (b) area of country, (c) projected population for 2050, (d) population mid-2009, (e) population density 2009, (f) projected population density 2050, (g) total increase in population density.

   (b) To calculate the population density for 2009 (column e) and 2050 (column f), you must do the following: divide the population mid-2009 and the projected population for 2050 by the surface area.

   Example: Israel (2009) = 7.2 million / 8131 square miles = 885.5 people per square mile

   Israel (2050) = 11 million / 8131 square miles = 1352.84 people per square mile

   (c) Now you can compare 2009 and 2050 projected population density values and fill in column g. We do this by calculating the percentage of change over time. This is done by setting up an equation to compare the two values. In the equation, the 2009 population density is taken to be 100%. We then calculate to see how much more dense it will be in 2015.

   Example: Israel = pop density 2009/100 = pop density 2050/x, solve for x.

   Israel = 885.5/100 = 1352.84/x

Israel = $(1382.54 \times 100)/885.5 = x$

Pop density percent change = 152.77%

Population density in 2050 will be 152.77% of the population density in 2009.

(d) Calculate the percentage of change for all the countries.

(e) Which countries are going to have the highest percentage of change in population density by 2050? Why do you think that is the case?

Review Exam

1. Describe the climatic map of North Africa and the Middle East (i.e., think about spatial patterns and spatial differentiation).

2. What impact did plant and animal domestication have on the peoples of the Fertile Crescent?

3. Define the term transhumance. How does transhumance differ from sedentary life?

____

____

____

____

____

4. Briefly describe the relationship between politics and population change in Israel-Palestine between 1890s and 1950s. How do the population dynamics of this region change between 1960 and the present?

____

____

____

____

____

5. What is the future of oil and water politics in the region?

____

____

____

____

____

6. The Red Sea has formed by which of the following processes?

   (a) collision/subduction
   (b) collision/continental overthrusting
   (c) horizontal displacement
   (d) seafloor spreading

7. Algiers is located in which of the following climatic regions?

   (a) tropical
   (b) dry arid/semiarid
   (c) midlatitude
   (d) highland

8. According to Iraq Body Count, the highest number of U.S. and UK military fatalities occurred in which period in Iraq?

   (a) May/June 2003
   (b) December/January 2003
   (c) November/December 2007
   (d) October/November 2004
   (e) March/April 2006

9. The sacred pilgrimage to Mecca, which is one of the Five Pillars of Islam, is called ____________________.

   (a) jihad
   (b) hajj
   (c) Islamism
   (d) alms giving

10. Which of the following can be identified as a cause of the genocide in Darfur
    (a) the move by Egypt to take over lands currently controlled by Sudan
    (b) the revolt of Islamic peoples against the native-centered African government
    (c) the revolt of native Africans against the Islamic-based government in Sudan
    (d) the desire of Arabs to settle the eastern region of Sudan near the Red Sea
11. Which of the following is the largest city in MENA?
    (a) Cairo (b) Khartoum
    (c) Ankara (d) Casablanca
12. Which of the following territories was a former Spanish colony?
    (a) Western Sahara (b) Algeria
    (c) Libya (d) Egypt
13. Which of the following are causes of outmigration from the region?
    (a) lack of economic opportunity (b) war (c) civil unrest
    (d) all of the above (e) none of the above
14. Islam is an Arabic term that means __________________.
    (a) submission (b) willingness to serve
    (c) desire (d) hope
15. The Maghreb has NOT been home to which of the following?
    (a) Romans (b) Vandals
    (c) Turks (d) Phoenicians

For each of the following statements, write T for "True" or F for "False" on the space provided.

16. Nearly three-quarters of all workers in the region are guest workers.__________________
17. The dominant form of Islam in Iran is Sunni Islam.__________________
18. Intifada is a term that means "uprising" and is used to represent the struggle of Palestinians seeking their own state within the occupied territories of Israel.__________________
19. Yemen remains one of the wealthiest oil-producing states in the Middle East.__________________
20. The headwaters of the Tigris and Euphrates Rivers are in modern-day Turkey.__________________
21. A chador is the sacred room in the mosque in which men pray.__________________
22. Most people living in squatter settlements (or *gecekondu*) in Istanbul, Turkey own land in rural areas of the country.__________________

23. Iran will have the lowest Inflation rate in MENA in 2010.____________________

24. Cotton is a major export out of Turkey.____________________

25. The Fertile Crescent is a historic region defined by the area within about 10 miles of the Nile River running from Sudan in the south to Egypt in the north.____________________

Mapping Key Places

Identify the following features on Map 4.1 (countries and cities) and Map 4.2 (landforms and waterways). On your political map, Map 4.1, label the major cities by writing the city names next to their respective symbols. When mapping some of the landforms, you may need to create some sort of symbol system that portrays the landform, such as a ^^^^^^^^^^^^ for a mountain range.

Countries	Major Cities	Landforms	Waterways
As listed in Chapter 4 of your textbook (Figure 4.1)	• Abu Dhabi • Algiers • Amman • Ankara • Aswan • Baghdad • Basra • Beirut • Cairo • Damascus • Doha • El Aalún • Fez • Istanbul • Jerusalem • Khartoum • Mecca • Medina • Riyadh • Tehran • Tel Aviv • Tunis • Tripoli	• Asir Mountains • Atlas Mountains • Elburz Mountains • Kopet Mountains • Pontic Mountains • Taurus and Anti-Taurus Mountains • Yemen Mountains • Zagros Mountains • Arabian Peninsula • Sinai Peninsula • Anatolian Plateau • Iranian Plateau • Arabian Desert • Libyan Desert • Nubian Desert • Sahara Desert	• Arabian Sea • Black Sea • Caspian Sea • Dead Sea • Mediterranean Sea • Red Sea • Persian Gulf • Gulf of Aden • Gulf of Sidra • Suez Canal • Euphrates River • Jordan River • Nile River • Tigris River • Straits of Hormuz

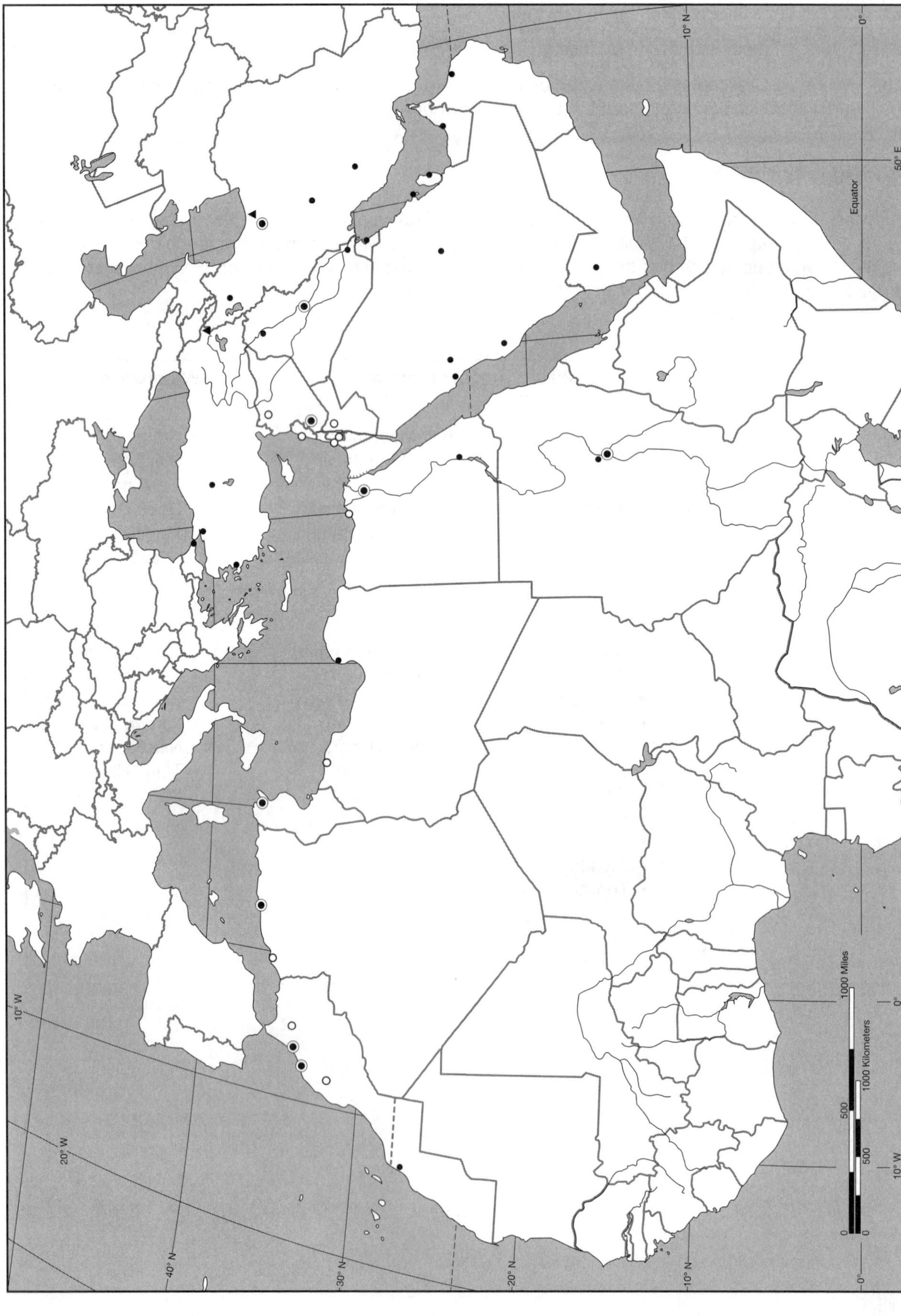

Map 4.1 Countries and Cities

Map 4.2 Landforms and Waterways

Mapping Internal Migration

Goal: In this exercise, you are going to map the countries in the region that have witnessed net immigration and net emigration.

1. Begin by reading the section in your text titled "Demography, Migration, and Settlement" (p. 132–137 and focus on Figures 4.25–4.28).
2. Based on that reading and your analysis of the key Figures in that section label each country on Figure 4.3 in the region as either being a country of high emigration "E" or a country of high immigration "I." For those countries that do not seem to have any data label them "ND." (Hint: Map 4.26 is particularly helpful.) Provide a key for your symbols on the map and include a title for both the key and map itself.
3. Which countries are net out-migration countries? Which are net in-migration countries?

4. What explains the difference between those countries that have higher levels of immigration and those countries that have higher levels of emigration?

5. How do the flows of international migration differ from the flows of intraregional migration? Where do most immigrants from the region migrate to internationally? What factors prompt those migrations?

6. The Palestinians have been forced to migrate more than other peoples in the region. What are some of the factors that have pushed Palestinians to migrate out of Palestine and Israel? In which countries can you find the most Palestinian migrants? What impact do these migrations have on their host countries?

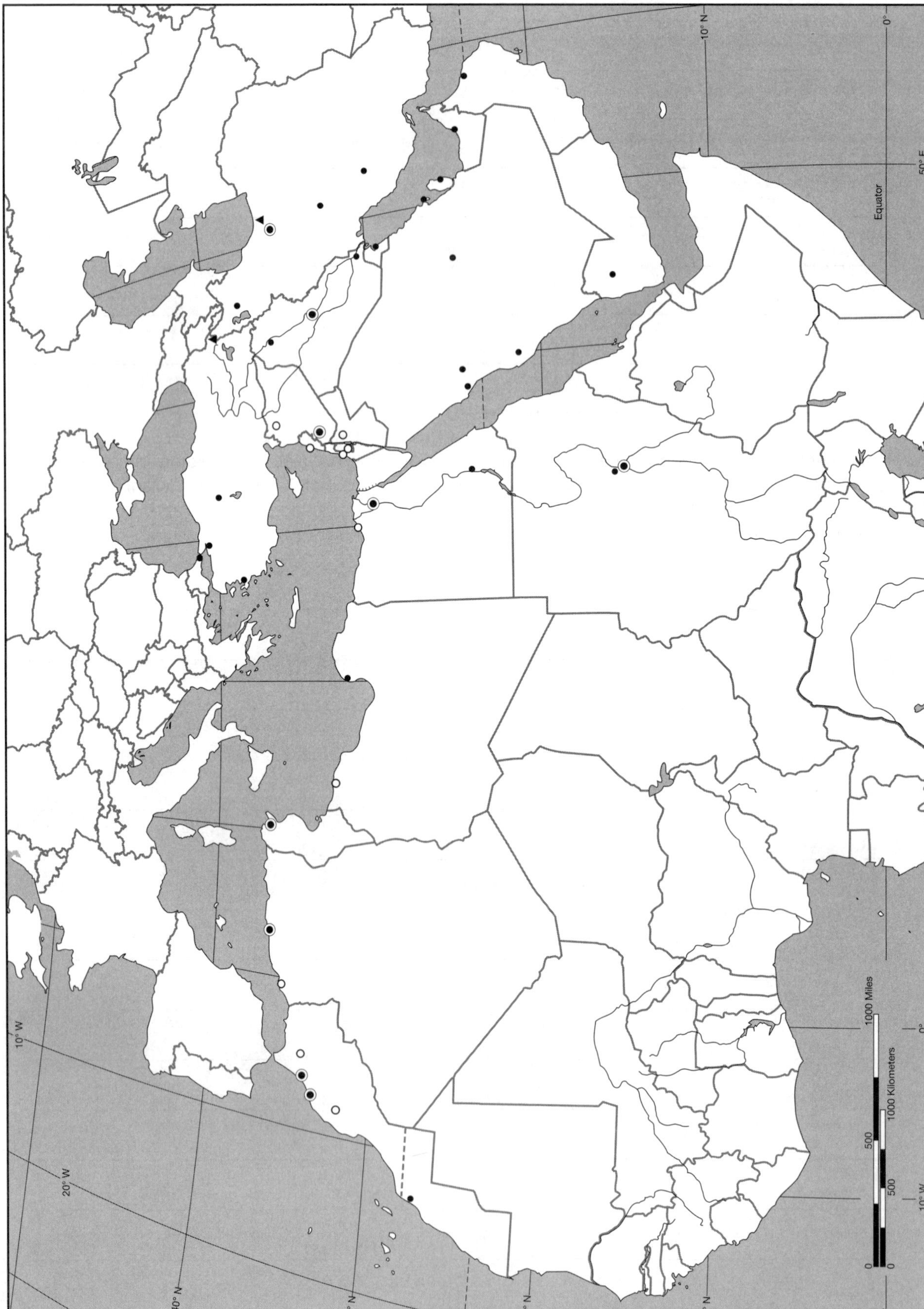
10° N
0°
50° E
Equator
10° W
20° W
40° N
30° N
20° N
10° N
0°
0° 0°
1000 Miles
1000 Kilometers
500
500
0
0
10° W

Map 4.3

CHAPTER 5

Sub-Saharan Africa

Learning Objectives

After reading and studying this chapter, you should be able to:

1. have an appreciation for the important historical position of this region in terms of human evolution and environmental change;
2. examine the impacts that global patterns of climate change will have on the region and its peoples over time;
3. understand the legacies of European colonialism and the impact of that colonialism and the population geography of this region;
4. comprehend the complex patterns of migration within and beyond this region and how these patterns are affected by various global and local social and political processes;
5. compare and contrast the complex socio-economic status of different peoples and sub-regions in sub-Saharan Africa, with a focus on the changing nature of economic development in the region; and,
6. appreciate the complexity of this region in terms of its physical and human landscapes.

Vocabulary Review

1. The winds, which are commonly called ____________________, emerge out of inland Africa carrying large amounts of dust and can cause very hot, dry conditions.

2. Plant ____________________ was an important part of the rise of civilizations in places such as Ethiopia, where peoples cultivated millet and other important crops thousands of years ago.

3. ____________________, which are not really red, are often used to purchase weapons because they are very high in value and easy to transport to international markets.

4. Today, the continent of Africa is made up of numerous countries. These modern-day political boundaries have their origin in Europe's ____________________, which was the rush to formally colonize all of the Africa initiated after the ____________________ of 1884–1885, when Otto von Bismark tried to negotiate European holdings on the African continent.

5. European influences on the continent have been numerous. In South Africa, the legacy of European colonization resulted in the establishment of a(n) ____________________ state under which blacks and whites were socially and spatially separated.

6. In the late 1950s and 1960s, the South African government put in place its strategy of spatial segregation through the development of ____________________, which were designed as "tribal reserves."

7. A recent report by the __________________ identified Africa as one of the most vulnerable regions to climate change. This term stands for ______________________________.

8. Sub-Saharan Africa is known for some of the highest levels of poverty in the world. Poverty, however, does not impact men and women equally. Women make up over two-thirds of the people living in poverty in the region. This process is known as the ____________________ and can be found in every region of the world today to different degrees.

9. In order to combat the differential effects of poverty on women, some countries have adopted the ____________________ approach designed by "Women in Development" to combat these

differences. The goals are to better link women's reproductive and productive roles and understand how the local social and spatial organization of society mediates women's experiences of development.

10. Following the successful implementation of the Grameen Bank in Bangladesh, some countries in Africa have also developed ___________________, whereby they provide small amounts of credit to families so that they can start their own businesses.
11. Somalia, located in a region known as ___________________ has seen some of the most severe political problems associated with their failed state.
12. ___________________ is only one form of shifting agriculture, whereby people clear land through burning to make room for domesticated staple crops.
13. ___________________, a modified version of Dutch, is a language that developed in South Africa among settlers who entered the region almost 400 years ago.
14. The theory that the physical geography of Africa is to blame for the region's underdevelopment echoes the theories of earlier geographers, who were known as ______________________.
15. Under British colonial rule, local leaders collected a ___________________ tax based on the total number of dwellings in the community. Unlike the French who ruled its colonies directly through assimilation, the British form of colonization was considered ___________________, and local leaders were cultivated and maintained tradition authority as a middle person between the British and the indigenous populations.
16. David Livingstone and Henry Stanley were both members of the ___________________, which was an organization founded for the "advancement of geographical science." This organization was tied closely to the colonial project in Britain.
17. The eradication of poverty and hunger as well as the promotion of gender equality and the empowerment of women are all part of a broader policy called the ___________________.
18. Trade languages, such as Swahili are often referred to as _________________.
19. While ________________ theorists often propose that development be aided by international assistance, ___________________ theorists suggested that underdevelopment was a legacy of colonialism.
20. It has long been argued that there exists an ________________ in sub-Saharan Africa, a legacy of colonialism and modern development practice, which both de-privilege rural areas and rural development.

Activities & Problems

1. Based on Figure 5.5, answer the following questions:
   (a) In which regions would you find an increasing spread of malaria?
   (b) In which regions are people most vulnerable to sea level rise?
   (c) Which region is currently not experiencing a food security problem?
2. Based on Figure 5.6, answer the following:
   (a) Which country has the largest proven oil reserve?___________________
   (b) Which country has the second largest proven gas reserve? ___________________
   (c) Name two countries that have proven oil reserves but no proven gas reserves.____________
3. Based on Figure 5.18, answer the following questions about colonization in Africa.
   (a) In what year was French West Africa created? ___________________ French Equitorial Africa? ___________________ Anglo-Egyptian Condominium? ___________________ Belgium Congo? ___________________
   (b) What were the names of Germany's colonial holdings? ___________________, ___________________, ___________________, ___________________

(c) What were the two Portuguese colonies named? ____________________,
____________________ The Portuguese could not connect their colonial holdings across the continent because the ____________________ were able to move north from their original holdings in southern Africa and claim the area between the two Portuguese holdings.

(d) The Chokwe's existing "state boundary" was divided between which colonial powers?
____________________, ____________________ And the Fulani? ____________________,
____________________, ____________________

4. Based on Figure 5.26, offer a brief discussion of the relationship between education and fertility rates in Mali, Ghana, and South Africa.

5. Examining Figure 5.3, describe the climatic patterns of the African continent. Using Figure 5.4 also briefly describe the map of high risk for desertification in the region.

6. Using Figure 5.24, offer a description of cell phone coverage in Africa. In which countries do we find the highest concentrations of cell phone users?

7. Perhaps even more acute than the North Africa and Middle East region is the problem of population density in Sub-Saharan Africa. In this question, you are going to: (a) calculate projected population densities for the countries of this region; (b) compare those calculations to the 2009

population density; and (c) examine the total percentage of change in population density over the two time periods.

(a) Turn to the "2009 World Population Data Sheet" located at the back of this text. On a separate piece of paper make the following columns in this order (you may want to use a spreadsheet program, such as Excel, for this exercise): (a) country name, (b) area of country, (c) projected population for 2050, (d) population mid-2009, (e) population density 2009, (f) projected population density 2050, (g) total increase in population density.

(b) To calculate the population density for 2009 (column e) and 2050 (column f), you must do the following: divide the population mid-2009 and the projected population for 2050 by the surface area.

Example: Tanzania (2009) = 37.9 million/364,900 square miles
= 104.12 people per square mile
Tanzania (2050) = 72.7 million/364,900 square miles
= 199.23 people per square mile

(c) Now you can compare 2009 and 2050 projected population density values and fill in column g. We do this by calculating the percentage of change over time. This is done by setting up an equation to compare the two values. In the equation, the 2009 population density is taken to be 100%. We then calculate to see how much more dense it will be in 2015.

Example: Tanzania = pop density 2009/100 = pop density 2050/x, solve for x.
Tanzania = 104.12/100 = 199.23/x
Tanzania = $(199.23 \times 100)/104.12 = x$
Pop density percent change = 191.35%
Population density in 2050 will be 191.35% of the population density in 2009.

(d) Calculate the percentage of change for all the countries.

(e) Which countries are going to have the highest percentage of change in population density by 2050? Why do you think that is the case?

___

___

___

___

___

___

___

___

___

___

Review Exam

1. Describe the landform geography of the Sub-Saharan region.

2. Describe the changing geography of AIDS cases in Africa between 1982 and 1997.

3. Why is Africa called the "cradle of mankind"? Explain.

4. What are some of the impacts that the Cold War had on the region?

5. In what way does the colonial legacy still affect the relationship between politics and peace in the region? Explain with examples.

6. Looking toward the future, what are some of the signs that economic change has occurred in this region? What are some of the challenges that remain?

7. Unlike southern Africa, which is dominated by Bantu-speaking peoples, the language family for Madagascar is ___________________.
   (a) Semetic-Hamatic
   (b) Nilotic
   (c) Malay-Polynesia
   (d) Afrikaaner

8. A person living in a Mediterranean climate zone in sub-Saharan Africa would live in which country?
   (a) South Africa
   (b) Namibia
   (c) Botswana
   (d) Lesotho

9. Which of the following is true about the colonial experience in Sub-Saharan Africa?
   (a) Some African groups actively resisted European rule by attacking European-controlled ports.
   (b) All African groups were pleased to have the advances offered by European colonizers.
   (c) European colonization did little to disrupt agricultural practices in the region.
   (d) Europeans were less interested in exploiting the mineral resources of the region than the previous ruling African kingdoms.

10. Which of the following following policies is reminiscent of Portuguese rule in Congo?
    (a) indirect colonial rule with the active participation of local populations in political decision-making
    (b) an active interest in promoting the assimilation of elites and the populace more generally

(c) ruthless control of land and labor with severe punishment for those who did not participate in public work projects

(d) an active investment in the social infrastructure of the local population

11. Which of the following languages are known as lingua franca or trade languages?

    (a) Hausa

    (b) Ibo

    (c) Zulu

    (d) Yoruba

12. Which country has NOT had UN intervention due to armed conflict?

    (a) Central African Republic

    (b) Angola

    (c) Zimbabwe

    (d) Côte d'Ivoire

13. Which of the following is most accurate?

    (a) The urban population of Sub-Saharan Africa in 2010 will be between 10 and 20 %.

    (b) The urban population of Sub-Saharan Africa in 2010 will be between 21 and 30 %.

    (c) The urban population of Sub-Saharan Africa in 2010 will be between 31 and 40 %.

    (d) The urban population of Sub-Saharan Africa in 2010 will be between 41 and 50 %.

14. If you were living in the region dominated by the Songhai Empire between C.E. 1400 and 1590, you would be living in which subregion in Sub-Saharan Africa?

    (a) West Africa

    (b) East Africa

    (c) Southern Africa

    (d) Central Africa

15. Programs that provide credit and savings to self-employed poor people who cannot get commercial loans so that they can start businesses on their own are often called ____________________.

    (a) sustainable development programs

    (b) social capital programs

    (c) indigenous technology programs

    (d) microfinance programs

For each of the following statements, write T for "True" or F for "False" on the space provided.

16. Diamonds are important commodities in Botswana.____________________

17. In January a large proportion of the African continent is impacted by dry, northeast trade winds.____________________

18. Malaria is found throughout the coastal areas of South Africa.____________________

19. Most places in Africa experienced colonization for hundreds of years.____________________

20. Apartheid policy included the Population Registration Act, which required all people in South Africa to be registered as either Bantu, colored, or white.____________________

21. There is no relationship between women's education and fertility in Mali. ____________________

22. Internal migration in Sub-Saharan Africa is a result of individuals seeking work and of refugees fleeing famine, floods, and violent conflict. ____________________

23. Coastal areas around the Nile Delta will be threatened by sea level rise due to climate change in the future. ____________________

24. The Bantu migrations took place from western Africa to southern Africa. ____________________

25. Despite increases in access to technology, there remains fairly stark social inequalities in the region. ____________________

Key Map Items

Identify the following features on Map 5.1 (countries and cities) and Map 5.2 (landforms and waterways). On your political map, Map 5.1, label the major cities by writing the city names next to their respective symbols. When mapping some of the landforms, you may need to create some sort of symbol system that portrays the landform, such as a ^^^^^^^^^^^^ for a mountain range.

Countries	Major Cities	Landforms	Waterways
As listed in Chapter 5 of your textbook (Figure 5.1)	• Abidjan • Adis Ababa • Brazzaville • Cape Town • Dakar • Dar es Salaam • Freetown • Harare • Johannesburg • Kinshasa • Lagos • Lusaka • Mogadishu • Monrovia • Nairobi • Timbuktu • Walvis Bay	• Chad Basin • Congo Basin • Djouf Basin • Kalahari Basin • Sudan Basin • Kalahari Desert • Namib Desert • Sahara Desert • Ethiopian Highlands • Great Rift Valley • Cape Fold Mountains • Ruwenzori Mtns. • Tibesti Mountains • Fouta Djallon Highlands • Horn of Africa	• Blue Nile River • Congo River • Niger River • Okavango River • Orange River • Senegal River • White Nile River • Zambezi River • Lake Chad • Lake Kariba • Lake Nyasa • Lake Tanganyika • Lake Victoria • Mozambique Channel • Gulf of Guinea

Map 5.1 Countries and Cities

Map 5.2 Landforms and Waterways

Mapping Colonial Legacies and Political Violence

Goal: The impact of European colonialism on this region cannot be underestimated. As the textbook highlights, one of the legacies of colonialism has been political violence. In this exercise, you are going to map out the pattern of European colonization at one point in time and layer onto that map a projection of Cold War Conflict and UN peacekeeping missions.You will need to read the textbook, particularly pages 166–172 and 188–191 and Figures 5.18, 5.20, and 5.33.

1. Using Map 5.3, begin by sketching out in different colors the rough colonial boundaries by 1914 based on the modern-day political map (these boundaries are not always perfectly aligned but they are fairly close, and note that Abyssinia was only under Italian protection until 1896 and was independent after that). Describe what pattern you see.

____

____

____

____

____

____

2. Place on the map a square for each "conflict and intervention" in the region since the Cold War began. Make sure to place the square within the boundaries of the appropriate country.
3. Place a circle for each "UN peacekeeping mission" into the region. Make sure to place the circle within the boundaries of the appropriate country.
4. Describe the pattern that you see. Is there any correlation between European colonial powers and modern-day conflict? Explain.

____

____

____

____

____

____

____

____

____

____

____

____

____

5. What impact did colonialism have on the modern-day political geography of Sub-Saharan Africa that may have facilitated the conflicts that you have mapped? Explain.

6. What regional organizations have emerged in the postcolonial period that might mitigate the affects of conflict in the region? If you were to map these supranational organizations onto the colonial map do you think you would see parallels between the two patterns? Explain.

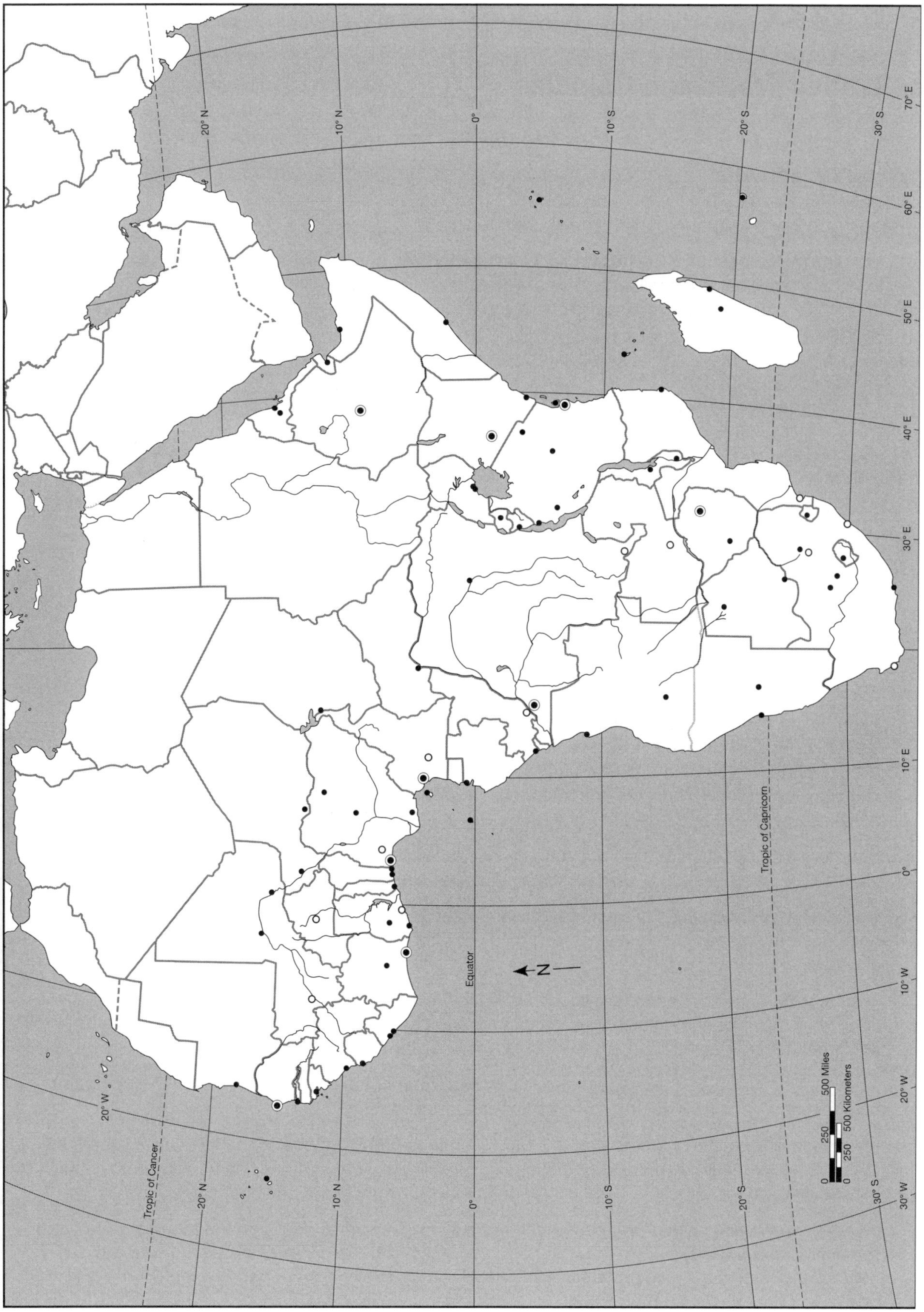
20° N
10° N
0°
10° S
20° S
30° S
70° E
60° E
50° E
40° E
30° E
10° E
0°
10° W
20° W
30° W
Equator
Tropic of Capricorn
Tropic of Cancer
N
0 250 500 Miles
0 250 500 Kilometers

Map 5.3

CHAPTER 6

The United States and Canada

Learning Objectives

After reading and studying this chapter, you should be able to:

1. examine the broader physical geography of this region and understand the ongoing concerns related to climate change and the vulnerabilities associated with that change as well as other natural hazards;
2. appreciate the complexity of this region in global context, particularity related to the experience of migration;
3. have a working knowledge of this region's economic development, understanding both its dynamic growth and its experience of socio-economic inequality;
4. consider the contributions that have come from this region to the wider cultural geography of other world regions; and,
5. compare and contrast the different economic, political, cultural and social practices found in the U.S. and Canada.

Vocabulary Review

1. Between the Pacific coastal ranges and the Rocky Mountains lies a(n) ____________________ region, which consists of a set of basins, plateaus, and smaller ranges.
2. As Europeans moved into North America they brought with them cultural, political, and economic systems. This process, called ____________________, was a result of French, British, Spanish, and Danish influences during the early colonial period.
3. Even before the United States became independent, people born and raised in the region were becoming less loyal to colonial powers. Through ____________________, a process by which people in the Americas took on a distinct cultural and political identity, people in the North American colonies defined a new culture based on liberalism, individualism, capitalism, and Protestantism.
4. Prior to the introduction of a slave economy into North America, ____________________ from places such as Britain were used as labor in the colonies.
5. Canada has developed its natural resources very effectively over the centuries and thus remains an important ____________________, providing unprocessed natural resources for many other countries.
6. The forced and self-imposed segregation of ethnic and racial groups in North America has sometimes fostered misunderstanding, which can result in ____________________, such as anti-immigrant or racial violence.
7. ____________________ are spaces of atmosphere that behave in the same way. These spaces can be measured and a value can be assigned to their quality.
8. In Canada, unlike the United States, the official state policy is not one of assimilation. Instead, the Canadian government is more supportive of ____________________, whereby individual cultural groups are allowed to assert their cultural uniqueness in a local and national context.
9. While North America has grown based on immigration, the overall population of the region has also remained quite mobile. ____________________, which is the process of people moving within a national territory, is quite common for many people living in North America, as people move from one state or province to another.

10. North America, more so than almost any other region in the world, has witnessed a high level of ____________________, as people continue to move to the fringes of urban areas.

11. In both the U.S. and Canada, one can find __________________________, which struggle against global authority centered in multinational corporations and international trade agreements.

12. Despite the decline in industrial labor, the changing capitalist system in North America has benefited from ____________________, whereby older industrial regions are "dismantled" in order to develop new, innovative enterprises and regions.

13. A sophisticated program, called ________________________, allows regional alliances to lower their emissions through government-imposed limits on carbon emission, emission permits, and exchange of permits between places.

14. The ___________________ is the total set of greenhouse gases caused by an individual, organization, event, product, or place.

15. In some areas, where industrial production developed largely unabated by legal sanctions, we now find ____________________, sites with extreme levels of pollution that have yet to be remedied.

16. A ____________________ is a region of extreme urbanization that has developed within the early industrial corridor of North America around Boston–New York–Washington, D.C., in the United States. In Canada, a similar corridor developed, called ____________________, between Windsor and Québec City.

17. _______________, which stands for ___________________________, is an organism that has had its DNA modified in a laboratory.

18. When agriculture is done without any commercial fertilizers, synthetic pesticides, or growth hormones, it is called ______________________.

19. Both Canada and the United States are ____________________, which means that power is allocated to local political units.

20. Some people seek to resist the move toward fast food by preserving cooking traditions tied to a newly defined ___________________ movement.

Activities & Problems

1. Using Figure 6.2 in your text and following the U.S./Canada border from the western coast of the United States to the eastern coast, trace the climate regions across the continent.

____

____

____

____

2. Examining Figures 2 and 3 from the Geographies of Indulgence, Desire, and Addiction box on page 204–205 in your text, answer the following:

   (a) List the top four tobacco exporting countries in the world today. ____________________, ____________________, ____________________, ____________________

   (b) List the top four tobacco-producing countries in the world today. ____________________, ____________________, ____________________, ____________________

(c) Which regions are projected to have the highest increase in smokers between 1998 and 2008?

____

(d) Which three countries will have the highest increase in the same time period?

__________, __________, __________

3. Using Figure 6.4, answer the following:

(a) Describe the pattern of total emissions of fossil fuel carbon for 2002 in the United States.

____

____

____

____

____

____

(b) Compare the pattern you see in Figure 6.4 with the map of population (Figure 6.18). What is the relationship between the pattern in Figure 6.4 and Figure 6.18?

____

____

____

____

____

____

4. Using Figure 6.17 describe or explain the following:

(a) Describe the pattern of poverty.

____

____

____

____

____

____

(b) Offer some explanation as to why the map looks the way it does.

____

5. Using Figure 6.19, answer the following:

(a) Describe the waves of immigration into the U.S. between 1800–2000.

(b) Currently, which region provides the largest number of migrants? Why?

(c) What factors caused the flow of migrants into the U.S. to slow down between 1930 and 1940?

Review Exam

1. Offer a brief explanation of how governors in the western United States might achieve the goal of reducing carbon emission.

2. Describe the pattern of subsistence practices of Native Americans in North America prior to European colonization.

3. Briefly describe and account for the most significant social movements in the United States and Canada.

4. In what ways has globalization opened up opportunities for Québec to establish itself as a more autonomous cultural region in Canada?

____

____

____

5. Briefly discuss the link between development in the "Pacific Rim" and what is often called "the new economy."

____

____

____

____

____

____

6. Which of the following is best supported by our current data/evidence on the earliest migration of humans to the North American continent?

   (a) The earliest any human appeared in North America was 11,000 years ago.

   (b) The earliest any human appeared in North America was 12,000 years ago.

   (c) The earliest any human appeared in North America was 15,000 years ago.

   (d) The earliest any human appeared in North America was 25,000 years ago.

7. Prior to 1492 it was possible that as many as ____________________ number of languages were spoken in North America.

   (a) 10 (b) 100 (c) 1000 (d) 2000

8. Immediately following colonization of North America by Europeans, the primary source of labor in the colonies was ____________________.

   (a) independent, free migrants from eastern Europe

   (b) slave labor from Africa

   (c) indentured servants from Europe

   (d) large groups of enslaved Native Americans

9. At the end of the 1800s the majority of migrants were coming from which world regions?

   (a) northern/western Europe and southern/eastern Europe

   (b) northern/western Europe and Asia

   (c) southern/eastern Europe and Asia

   (d) Latin America and Asia

10. Which of the following is true of the differences between migration patterns to the United States and Canada during the colonial period?

    (a) Most people migrating to Canada were from eastern Europe.

    (b) Most people migrating to Canada were British and French.

    (c) Most people migrating to the United States were British and French.

    (d) Most people migrating to the United States were German.

11. Which of the following has NOT had a significant impact on the process of suburbanization in North America?

(a) horse-drawn streetcars (b) rail services

(c) automobiles (d) airplanes

12. Which of the following is representative of policy put in place by the Canadian government to ward off the invasion of American culture?

(a) limiting the number of Americans that can cross into Canada each year

(b) levying a tax on *Sports Illustrated* because it did not contain enough material on Canadian sports

(c) building a large satellite system that can stop the flow of U.S. television into Canada

(d) mandating that 100% of all music on Canadian radio must be Canadian

13. Movements that challenge corporate control over the economy or more covertly challenge government control are sometimes referred to as ________________________.

(a) cap-and-trade movements (b) antiglobalization movement

(c) conspiracy movements (d) anti-colonial movements

14. The Pacific Rim refers to which of the following?

(a) all countries in the Americas that border the Pacific

(b) all countries that border the Pacific

(c) all countries in Asia that border the Pacific

(d) the United States, Japan, and China

15. Which of the following major Canadian cities is not located in the area called Main Street?

(a) Ottawa (b) Montreal (c) Québec (d) Vancouver

For each of the following statements, write T for "True" or F for "False" on the space provided.

16. The United States and Canada are both members of the international Asia-Pacific Economic Cooperation organization.____________________

17. The North American core in the United States is called Main Street.____________________

18. The Monroe Doctrine encouraged Europeans to increase their political and economic authority in the western hemisphere.____________________

19. The slow food movement is about eating animals that are often sedentary or non-moving.

____

20. A 1/4 pound of beef hamburger imported from Latin America requires the clearing of 6 square yards of rain forest. ____________________

21. Government policy in Canada has been supportive of monoculturalism, based on the philosophy that everyone should subscribe to one set of cultural values that is Canadian.____________________

22. The first major wave of internal migration in the region was marked by the movement of people of European descent from the eastern seaboard into the interior parts of the continent.____________________

23. The new regional economy of North America is marked by a recent and steep rise in industrialization.____________________

24. The Canadian government has been more likely, historically, to guarantee social welfare to its citizens than the United States.____________________

25. Most fossil fuel carbon emission takes place in states such as Nevada and Montana. ____________________

Key Map Items

Identify the following features on Map 6.1 (countries and cities) and Map 6.2 (landforms and waterways). On your political map, Map 6.1, label the major cities by writing the city names next to their respective symbols. When mapping some of the landforms, you may need to create some sort of symbol system that portrays the landform, such as a for a ^^^^^^^^^^^^ mountain range.

Countries	Major Cities	Landforms	Waterways
As listed in Chapter 6 of your textbook (Figure 6.1)	• Anchorage • Atlanta • Boston • Calgary • Chicago • Detroit • Edmonton • Kansas City • Las Vegas • Los Angeles • Miami • Minneapolis-St. Paul • Montréal • New Orleans • New York • Ottawa • Philadelphia • Phoenix • Portland • Québec • Sacramento • Salt Lake City • San Antonio • San Diego • San Francisco • Seattle • St. Louis • Toronto • Vancouver • Washington, D.C. • Winnipeg	• Adirondack Mtns. • Alaska Range • Appalachian Mtns. • Baffin Island • Brooks Range • Cascade Range • Coast Mountains • Colorado Plateau • Grand Canyon • Greenland • Mojave Desert • Newfoundland • Ozark Mountains • Piedmont • Rocky Mountains • Sierra Nevada Mtns. • Teton Range	• Arkansas River • Colorado River • Columbia River • Mackenzie River • Mississippi River • Missouri River • Ohio River • Platte River • Red Rive • Rio Grande River • St. Lawrence River • Gulf of Alaska • Gulf of Mexico • Baffin Bay • Chesapeake Bay • Hudson Bay • Great Lakes • Great Salt Lake • Great Slave Lake • Lake Winnipeg • Bering Strait

Map 6.1 Countries and Cities

Map 6.2 Landforms and Waterways

Mapping Poverty Rates

Goal: In this exercise, you are going to examine the patterns of poverty across the United States based on an average of poverty data at the state level from 2006–2008.

1. You are going to begin by organizing the following data into five statistical groups from those with the highest rate of poverty to those with the lowest rate.

State	Percent*	State	Percent*	State	Percent*
Alabama	14.4	Kentucky	16.5	N. Dakota	10.8
Alaska	8.2	Louisiana	17.1	Ohio	12.9
Arizona	15.6	Maine	0	Oklahoma	14.1
Arkansas	15.6	Maryland	8.6	Oregon	11.7
California	13.2	Mass.	11.5	Pennsylvania	20.9
Colorado	10.2	Michigan	2.4	Rhode Island	20.9
Connecticut	8.3	Minnesota	9.1	S. Carolina	13.1
Delaware	9.4	Mississippi	20.5	Tennessee	24.9
D.C. (Wash.)	17.6	Missouri	2.5	Texas	26.3
Florida	12.4	Montana	13.1	Utah	8.8
Georgia	3.9	Nebraska	10.2	Vermont	8.9
Hawaii	8.9	Nevada	10.0	Virginia	9.2
Idaho	210.6	New Hamp.	6.1	Washington	9.5
Illinois	0	New Jersey	8.9	W. Virginia	14.9
Indiana	12.3	New Mexico	16.7	Wisconsin	10.3
Iowa	9.6	New York	14.2	Wyoming	10.3
Kansas	12.4	N. Carolina	14.4		

* These data are from U.S. Census Bureau, Current Population Survey, 2006, 2007, and 2008 U.S. Census Bureau, Poverty Statistics http://www.census.gov/hhes/www/poverty/poverty08/tables08.html

2. Rank the states listed above from places of highest poverty to places of lowest poverty. Create five categories based on these statistics; and provide shading that corresponds to each of these categories using your colored pencils.
3. Shade Map 6.3 accordingly, providing a key and title for the map to illustrate the extent of each data category.
4. Describe the pattern that you see. Is there any clustering of states based on their standing? If so, how are they clustered?

____

____

____

____

____

____

____

5. Do you see any regional clustering of poverty? Are there any states that are outliers relative to their neighboring states? Is there any consistent spatial pattern within the cluster of states with the highest rates? Why or why not?

N
Arctic Circle
Tropic of Cancer
60° N
70° N
50° N
40° N
30° N
20° N
22° N
20° N
20° W
30° W
40° W
50° W
60° W
0° W
90°
120° W
140° W
150° W
160° W
170° W
180°
156° W
158° W
160° W
0 250 500 Miles
0 250 500 Kilometers
0 75 150 Miles
0 75 150 Kilometers

Map 6.3

CHAPTER 7

Latin America and the Caribbean

Learning Objectives

After reading and studying this chapter, you should be able to:

1. comprehend the complex physical geographies of this region and how people have adapted to that complexity;
2. appreciate the importance that various migratory populations have had in this region, as people have been brought to the region from Europe, Africa, and Asia;
3. appreciate the complex historical geographies of this region from the pre-Columbian through the periods of European colonialism and independence;
4. compare and contrast the dynamic economic and political geographies of this region, paying particular attention to how those geographies affect the demography of the region; and,
5. have a working knowledge of the cultural geography of the region as well as how that geography is related to issues of indigenous rights as well as violence and corruption in the region.

Vocabulary Review

1. ____________________ can be used to classify Latin America's mountain regions into four distinct climatic, vegetative, and human activity subregions.
2. The ____________________ is a theory that the Americas were untouched prior to European contact in 1492 and that native peoples lived in harmony with their environment.
3. In 1494, Pope Alexander VI drew a line of demarcation under the ____________________, at which time the world was divided between Spanish and Portuguese interests.
4. The term ____________________ denotes the process by which native populations in the Americas were decimated by diseases brought by Europeans.
5. Colonial powers in Latin America and the Caribbean established ____________________, large estates through which they raised cattle, wheat, olives, and fruit for sale in Europe.
6. ____________________ took the place of indigenous agricultural systems and were used to organize the production of single crops, such as sugar.
7. The plateau region located in the Andes is often called the ____________________.
8. In the 1980s and 1990s, the economies of Latin America and the Caribbean became quite sluggish and some witnessed stagflation (a combination of economic stagnation and hyperinflation). In the wake of these economic crises, the International Monetary Fund imposed ____________________ under which governments were to reduce spending on subsidies and trade barriers as well as privatize government-owned industries in exchange for large-scale loans.
9. ____________________ is a term that was used to describe someone of Spanish/African descent during the colonial period, while ____________________ was used to describe someone of African/Indian descent.
10. In countries where one city dominates political, economic, and social life, we often call that ____________________.
11. ____________________ were given work permits to enter the United States from Mexico between 1942 and 1964. Many of these migrants never returned to Mexico.
12. A new form of Catholic practice, known as ____________________, focuses attention on the poor and disadvantaged segments of society. Practitioners mixed the ideas of Jesus Christ with the philosophies of Karl Marx.

13. In Brazil, the informal settlements that emerge around cities are called ____________________. These spaces are homes to millions of poor migrants.
14. Many governments have moved away from growing and selling grains on the international market, instead focusing on ____________________, such as fruits, vegetables, and flowers.
15. People who do not work in waged jobs that are often taxed, might instead participate in the ____________________, which includes a variety of self-employed, income-generating activities.
16. The ____________________ was defined by historian Alfred Crosby to explain the interchange of crops, animals, and diseases between the Americas, Europe and Africa.
17. A fairly recent form of economic development called ____________________ is intended to both provide local people with jobs while simultaneously protecting endangered ecosystems.
18. The hunt for new and unique tropical flora, known as ____________________, is driven by the search for new medicines and other commercial uses of rare plant life.
19. In 1823, the United States established the ____________________, which stated that any further European colonization or interference in the Western Hemisphere would be seen as an act of aggression.
20. Banks in the Bahamas and the Cayman islands offer ____________________, including tax-exempt and confidential banking.
21. ____________________ are monies that are sent home by migrants often working in either another country or in a large domestic urban industrial area. This income is essential to many families globally, as more and more people migrate to earn income.
22. Throughout the Caribbean, one can find ____________________ where goods are manufactured without any custom duties.
23. Many multinational corporations, such as Del Monte, use a strategy called ____________________ by which local farmers sign contracts to produce crops for certain production and at certain quality standards for a guaranteed price.
24. The term ____________________ refers to the process by which pesticides are imported into Latin America, used on crops, and those crops are then returned to the country that produced the pesticide.

Activities & Problems

1. Using Figure 7.3, trace the different climatic regions across South America at the 20th southern parallel.

____

____

2. Using Figure 7.4, list three primary commodities that one could find in the following altitudinal zones:

   (a) Tierra Caliente: ____________________ ____________________ ____________________

   (b) Tierra Templada: ____________________ ____________________ ____________________

   (c) Tierra Fria: ____________________ ____________________ ____________________

   (d) Tierra Helada: ____________________ ____________________ ____________________

3. Based on Figure 7.32, name the year in which the U.S. intervened in the following countries.

   (a) El Salvador ____________________

   (b) Chile ____________________

   (c) Argentina ____________________

   (d) Grenada ____________________

   (e) Guatemala ____________________

4. Based on Figure 7.29, name the language that is most commonly spoken in the following places:

   (a) the east coast of Brazil ____________________

   (b) northern Mexico near the U.S. border____________________

   (c) Haiti ____________________

   (d) Bahamas ____________________

   (e) Suriname ____________________

   (f) Guatemala ____________________

5. Using Figure 7.14 and the accompanying text, describe the geographic extent of the Maya, Aztec, and Incan empires. Name some of the major accomplishments of these early world historical societies.

6. Using Figures 7.2 through 7.13 and the accompanying text, respond to the following statements and questions related to the physical and resource geography of Latin America and the Caribbean.

   (a) Describe the broad physical regions and landforms of the region.

   (b) Compare and contrast the varied landscapes that are a result of the complex relationship between landform and climatic zones in the region.

(c) Discuss the relationship between landform geography and mineral resource industries in the region.

(d) Explain the impacts of El Niño and La Niña on the region.

(e) What is meant by the term biodiversity? Why is this region such a rich source of biodiversity?

Review Exam

1. What are some of the reasons given for the collapse of Mayan civilization between C.E. 500 and C.E. 1000?

2. How did the Incas adapt to their environmental conditions?

3. Compare and contrast the stark differences between rich and poor in the region today.

4. What is liberation theology? What sparked this movement and what resulted from its inception?

5. Briefly examine the relationship between the geological history, the region, and its mineral wealth.

6. Briefly describe the future of economic relations in the region. How does this future relate to the future of human-environment relations in the region?

7. If you were living on the coast of Peru in Lima, you would most likely experience what type of climatic region?

   (a) tropical rainforest
   (b) tropical or subtropical desert
   (c) tropical savanna and grassland
   (d) midlatitude grassland

8. If you were in Brasilia, the capital of Brazil, you would most likely experience what type of climatic region?

   (a) tropical rainforest
   (b) tropical or subtropical desert
   (c) tropical savanna and grassland
   (d) midlatitude grassland

9. Despite interest by Spanish and Portuguese colonial powers to exploit local labor, efforts to use local labor were frustrated by the rapid decline of indigenous populations. This process is known as ____________________.

   (a) demographic explosion
   (b) demographic exchange
   (c) demographic collapse
   (d) demographic exposure

10. Which of the following is true about the exchange between Europe and Latin America and the Caribbean?

    (a) The Europeans introduced rats, pigs, and potatoes to Latin America.
    (b) The Europeans introduced rats, pigs, and corn to Latin America.
    (c) The Europeans moved corn and cocoa from the Americas to Africa.
    (d) The Europeans moved pigs and rats from the Americas to Africa.

11. Which of the following countries received the largest amount of slave labor from Africa?

    (a) Brazil
    (b) Cuba
    (c) United States
    (d) Argentina

12. Which of the following are you likely to find in the Teirra Helada?

    (a) poultry
    (b) Vicuna
    (c) sugarcane
    (d) EU

13. During the colonial period, if you lived in Latin America or the Caribbean but you were born in Spain, you would be called which of the following?

    (a) criollos
    (b) mulatto
    (c) peninsulares
    (d) zambo

14. The Caribbean Development Initiative (CDI) is a ____________________.

    (a) free trade association
    (b) command economy association
    (c) nontraditional economy cooperative
    (d) an indigenous commodities production program

15. A speaker of Quechua, a language indigenous to South America, would most likely live in which region?

    (a) central South America region
    (b) Caribbean region
    (c) central Mexico region
    (d) Andean region

16. Which of the following is true about people who live in the Andes?

    (a) Most people grow crops for the international market.
    (b) Most people grow potatoes close to the base of the mountains.
    (c) Most people grow agricultural products for subsistence or local consumption.
    (d) Most people use horses to move goods up and down steep mountain paths.

For each of the following statements, write T for "True" or F for "False" on the space provided.

17. Caribbean culture is strongly influenced by African cultural practices. ____________________

18. A farmer growing potatoes in the Andes would have to grow this crop in the Tierra Fria or Tierra Helada altitudinal zones. ____________________

19. The term Latin America was coined by the Portuguese in the 17th century. ____________________

20. Most of the mineral wealth in Latin America is found in the old crystalline rock where crustal folds bring older rock near the surface. ____________________

21. Haiti was declared independent in 1804, when slaves inspired by the French Revolution led a coup against their former slaveholders. ____________________

22. World War I had a positive effect on Latin America's industrial development because of increasing demands for needed raw materials from the region. ____________________

23. The Latin American and Caribbean region make up approximately 5% of the world's overall GNP, which is higher than Africa, South and Southeast Asia, the Middle East, East Asia, and the Pacific. ____________________

24. Most braceros that migrated to the United States in the 1960s and 1970s quickly returned to Mexico after earning money for their families. ____________________

25. Gender inequality is minimal in the region, with women's and men's literacy rates largely equal throughout the region. ____________________

Key Map Items

Identify the following features on Map 7.1 (countries and cities) and Map 7.2 (landforms and waterways). On your political map, Map 7.1, label the major cities by writing the city names next to their respective symbols. When mapping some of the landforms, you may need to create some sort of symbol system that portrays the landform, such as a ^^^^^^^^^^^^ for a mountain range.

Countries	**Major Cities**	**Landforms**	**Waterways**
As listed in Chapter 7 of your textbook (Figure 7.1)	• Bogotá • Brasilia • Buenos Aires • Caracas • Cuidad Juárez • Guatemala City • Havana • Kingston • La Paz • Lima • Mexico City • Panama City • Port-au-Prince • Rio de Janeiro • San Juan • Santiago • Santo Domingo • São Paulo • Tijuana	• Amazon Basin • Andes Mountains • Atacama Desert • Brazilian Highlands • Greater Antilles • Guiana Highlands • Lesser Antilles • Pampas • Patagonia • Sierra Madre Occidental Mtns. • Sierra Madre Oriental Mtns. • Sonoran Desert • Yucatan Peninsula	• Amazon River • Caribbean Sea • Gulf of California • Gulf of Mexico • Lake Nicaragua • Lake Titicaca • Orinoco River • Paraná River • Rio de la Plata • Rio Grande River • Uruguay River

Map 7.1 Countries and Cities

Map 7.2 Landforms and Waterways

Mapping Ecosystems

Goal: In this exercise, you are going to investigate the relationship between landform geography and the climatic patterns of Latin America. For this exercise you will need to use your textbook (particularly Maps 7.3 and 7.7, but also 7.5) as a source of data for your map.

1. You are going to begin by placing on Map 7.3 all the major physiographic features (landforms) from the table above (use one color to represent these various landforms, using the symbol ^^^^^^^^^^ for mountains, for example). Make sure you get some sense of elevation for your map.
2. On the same map, trace out, using distinct colors, the major climatic regions.
3. Describe the climatic pattern that you see.

____

____

____

____

____

____

____

____

____

4. In what sense do you believe that the climatic patterns you see on this map are a reflection of the physiographic features of this particular region? Explain.

____

____

____

____

____

____

____

____

____

____

____

5. What other factors, beyond the physiographic, help explain the climatic map of this region? Explain. Try to represent these processes on your map.

60° W
50° W
40° W
30° W
40° N
30° N
Tropic of Cancer
20° N
10° N
Equator
0°
10° S
N
20° S
Tropic of Capricorn
30° S
40° S
0 200 400 Miles
0 200 400 Kilometers
50° S
130° W
120° W
110° W
100° W
90° W
80° W
70° W
60° W
50° W
40° W
30° W
20° W
60° S

Map 7.3

CHAPTER 8

East Asia

Learning Objectives

After reading and studying this chapter, you should be able to:

1. have a working knowledge of the dynamic physical geographic differences present in the region;
2. understand how this region's physical geography has been massively restructured by human intervention over time;
3. appreciate that while the longer-term historical geography of this region is dominated by mainland China, in more recent centuries Japan and Taiwan have been important political players in the region, with South Korea also playing an important role in the region's post-Cold War politics;
4. explore the divergent economic development paths that the countries in this region have taken in the post-World War II period;
5. compare and contrast the diverse sets of cultural practices and landscapes at work in this region, which are themselves a result of intra-regional and global relations; and,
6. understand the current geopolitical concerns that mark this region and comprehend the possible concerns that may present themselves in the near future.

Vocabulary Review

1. In the post-1868 period in Japan, the government supported capitalist monopolies, called ____________________, through public funding.
2. The ____________________ is a loosely defined region of countries that border the Pacific Ocean.
3. The Japanese archipelago is a series of volcanic islands that make up part of the larger ____________________, which girdles the Pacific Ocean.
4. ____________________, what can be broadly defined as animated film, is actually its own genre, with characteristics that can be both real and fantastic.
5. ____________________ are business organizations that were established in Japan as part of post-WWII economic reforms.
6. The ____________________ include the countries of Japan, South Korea, Taiwan, and Singapore, a number of newly industrialized countries that have witnessed rapid economic growth.
7. In the 1970s, there was a brief period of ____________________, when businesses and people decided to escape congestion and inflated land prices in metropolitan Japan.
8. The philosophy that one can analyze the physical attributes of places and improve the flow of cosmic energy is called ____________________ in Chinese, something that is practiced widely in China.
9. The early empires of East Asia are sometimes referred to as ____________________ because they were based on the ability and capacity of these societies to control water, particularly rivers.
10. Under Mao Zedong, the Chinese set out to expand economic development through a radical restructuring of land ownership and use. This policy, called the ____________________ actually led to massive famine in the country.
11. The South Korean government facilitated capitalist development by supporting ____________________, a group of industrial conglomerates.
12. The Tibetan Plateau is a ____________________, a mountainous block of earth that is surrounded by faults and folds, creating a displaced unit of uplifted rock.

13. The Chinese language is a difficult one to learn, but the government has tried to simplify the language by reducing the number of characters used as well as adopting a new system, called ____________________, for spelling Chinese words and names using the Latin alphabet.

14. Chinese writing, which dates back almost 4000 years, is based on ______________________, each of which represents a picture or idea.

15. ___________________ is a distinctly Japanese religious form based in the notion that every individual thing can be embued with sacred power.

16. Beginning in the late 1970s, China embarked on an economic plan that included the development of __________________________ or ________ for short. These spaces were purposively segregated from the rest of Chinese society.

17. In South Korea, the post-World War II economic model was based on the concept of ____________________, whereby domestic industries would be protected from external competition.

18. China's goals of reducing its overall population growth are epitomized in the ______________________, which has strictly regulated fertility rates in the country.

19. ______________________ is more a philosophy than a religion, as it makes no pronouncements about gods or the afterlife.

20. Cartoon books, popularly known as ___________________, are important cultural artifacts in Japan today. In fact, the production of these books is an international phenomenon today.

Activities & Problems

1. Using Figure 8.3, answer the following questions:

   (a) What kind of climate would you find in Ulan Batar? ____________________

   (b) What kind of climate would you find in Shenyang? ____________________

   (c) What kind of climate would you find in Guangzhou? ____________________

   (d) What kind of climate would you find in Pyongyang? ____________________

   (e) What kind of climate would you find on the island of Hainan? ____________________

   (f) What kind of climate would you find in China in the region bordering South Asia?

   ____

2. Using Figure 8.15 in your text, describe the following.

   (a) Describe the extent of the Japanese Empire just prior to 1931.

   ____

   ____

   ____

   ____

   ____

   ____

   (b) Describe the extent of the Japanese Empire after 1931 and before 1933.

   ____

   ____

(c) Describe the expansion of the Japanese Empire in 1933.

3. Using Figure 8.18 in your text, describe the pattern of population density in East Asia in 1995. What impact might the spatial distribution of population have on resource allocation in the region?

4. Using Figure 8.21, answer the following questions:

   (a) Where would we find the largest Chinese immigrant community outside of China in the world today? ____________________

   (b) From which places have 19th century Chinese immigrant communities largely vanished today?

   ____

   (c) Which two countries have been the largest recipients of Chinese immigration in Europe?
   ____________________ and ____________________

5. Human-environment interactions is one of the main areas of study in geography. As a budding geographer, you need to consider how humans have adapted to the physical environment, modifying that environment to their needs. To do this, you also need to understand the different environmental contexts in which people live. With this in mind, answer the following:

   (a) Using Figures 8.3 and 8.5, and the text in the section entitled "Ecology, Land, and Environmental Management" describe the differing climatic vegetation, and physiogeographic subregions of East Asia.

(b) Explain what primary economic activities you are likely to find in each of the regions you described in 5(a).

(c) Describe the relationship between primary economic activity and population density one might find in each of these areas across the region based on your answers in (a) and (b).

(d) Using Figure 8.7 in your text, do the following: (a) describe what you see; and (b) explain the relationship between the physical and human geographies in this particular place as well as how the spatial organization of this place might impact local social organization.

(e) Using Figure 8.8 in your text, do the following: (a) describe what you see; and (b) explain the relationship between the physical and human geographies in this particular place as well as how the spatial organization of this place might impact local social organization.

Review Exam

1. Name two of the most significant contributions of the Qin dynasty to classical world China.

2. Describe the Asian Brown Cloud.

3. In what two areas do we find the highest level of ethnic tension in China? What is the main cause of these tensions?

4. What is the "Great Leap Forward" and what impact did it have on Chinese society?

5. What impact has the liberalization of the Chinese economy had on rural-to-urban migration in China?

6. The Shang dynasty was responsible for which of the following?
   (a) establishing the first organized territorial state in China
   (b) developing civilization in southern China
   (c) building the first walled cities in China
   (d) subdividing China into several smaller empires

7. In which of the following regions would one most likely find nomads today?
   (a) Japan's Pacific corridor
   (b) South China
   (c) Manchuria
   (d) Mongolia

8. Which of the following is true of good in East Asia?
   (a) Food traditions in East Asia are both regionally distinct and a product of East Asia's relationship with other parts of the world.
   (b) All food in East Asia is a result of the relationship of this region to other parts of the world.
   (c) The foods of East Asia are uniformly similar regardless of where one might find one's self.
   (d) The meat-filled dumpling was most likely created in Outer China in the 10th century B.C.E.

9. Which of the following is not true of Japan's postwar economic miracle?
   (a) There were exceptionally high levels of personal savings in Japan.
   (b) Japan was able to acquire new technologies from the United States.
   (c) The Japanese government fostered local industry through a series of tax concessions.
   (d) The Japanese relied heavily on their vast body of natural resources, such as coal and oil, to foster the growing industrial economy.
10. Which of the following is true of Xinjiang, a territory that occupies one-sixth of China?
   (a) The majority of the population is Christian.
   (b) The majority of the population is Muslim.
   (c) The majority of the population is Buddhist.
   (d) The majority of the population is animist.
11. Which of the following is true of Taiwan?
   (a) The Nationalist government set up the Republic of China on the island after being forced out of the mainland by the Communists.
   (b) The Nationalist government established a democratic regime almost immediately in Taiwan.
   (c) Taiwan gained full diplomatic status in 1971 after Nixon visited the People's Republic of China.
   (d) Taiwan's economic growth rate has stagnated since the 1970s.
12. What is meant by the term *juche*?
   (a) This is a term used in South Korea and represents a mix of Stalinist socialism, self-reliant nationalism, and a cult of personality.
   (b) This is a term used in North Korea and represents a mix of Stalinist socialism, self-reliant nationalism, and a cult of personality.
   (c) This is a term used in North Korea and represents a mix of state capitalism, self-reliant nationalism, and a cult of personality.
   (d) This is a term used in South Korea and represents a mix of state capitalism, self-reliant nationalism, and a cult of personality.
13. In which area of East Asia are you most likely to find wet rice agriculture?
   (a) the Tibetan Plateau
   (b) the central mountains and plateaus region
   (c) the southern continental margin
   (d) the northern continental margin
14. Which of the following is true of Japan's imperial period?
   (a) They successfully conquered the region around Shanghai in southern China.
   (b) They successfully conquered South but not North Korea.
   (c) They successfully conquered North but not South Korea.
   (d) They successfully conquered both North and South Korea.
15. Which of the following best defines Confucianism?
   (a) This philosophy was based on worshiping numerous gods.
   (b) This philosophy was based on ethics and the principles of matriarchy.
   (c) This philosophy was based on ethics and the principles of good governance.
   (d) This philosophy was based on family and matrilineal descent.

For each of the following statements, write T for "True" or F for "False" on the space provided.

16. The wealth of the Chinese has been quite limited during the most recent period of economic growth in the country. ____________________

17. The Chinese were successful in repelling the British during the Opium War of 1839–1842. ____________________

18. In the early years of the People's Republic of China Mao Zedong wanted a large population to exploit China's large territory. ____________________

19. Japan has witnessed a growing trend where women in their mid-to late 20s refuse to marry; instead they live at home and spend their disposable income on luxury items. ____________________

20. Pastoralism and nomadism has been the basis of life on the North China Plain for almost 5,000 years. ____________________

21. Throughout mainland Asia rivers carry massive amounts of silt, which is deposited in the flatter lowland areas of the region. ____________________

22. Kim Jong Il, the son of Kim Il Sung, is known by his people as "Great Leader." ____________________

23. Taiwan is considered to be one of the Asian Tigers along with Japan, North Korea, and Singapore. ____________________

24. There has been an attempt recently to reintroduce *tu long* into the wild in China. ____________________

25. In recent years, North Korea has witnessed an explosion of food production that has led to food surpluses for the last five years. ____________________

Key Map Items

Identify the following features on Map 8.1 (countries and cities) and Map 8.2 (landforms and waterways). On your political map, Map 8.1, label the major cities by writing the city names next to their respective symbols. When mapping some of the landforms, you may need to create some sort of symbol system that portrays the landform, such as a ^^^^^^^^^^^^ for a mountain range.

Countries	Major Cities	Landforms	Waterways
As listed in Chapter 8 of your textbook (Figure 8.1)	• Biejing • Guangzhou • Hong Kong • Kyoto • Osaka • Pyongyang • Seoul • Shanghai • Tokyo • Ulan Bator • Xi'an	• Altay Mountains • Daba Range • Hainan Island • Himalaya Mountains • Khingan Range • Kunlun Mountains • Mongolian Plateau • Pamirs Mountains • Qinling Range • Sayan Mountains • Sichuan Depression • Tarim Basin • Tibet Plateau • Tien Shan Mtns. • Turfan Depression • Yablonovyy Range • Yunnan-Guizhou Plateau • Zunghaer Basin	• Bo Hai Bay • Chiang Jiang River • East China Sea • Huang He River • Mekong River • Sea of Japan • South China Sea • Xi River • Yellow Sea

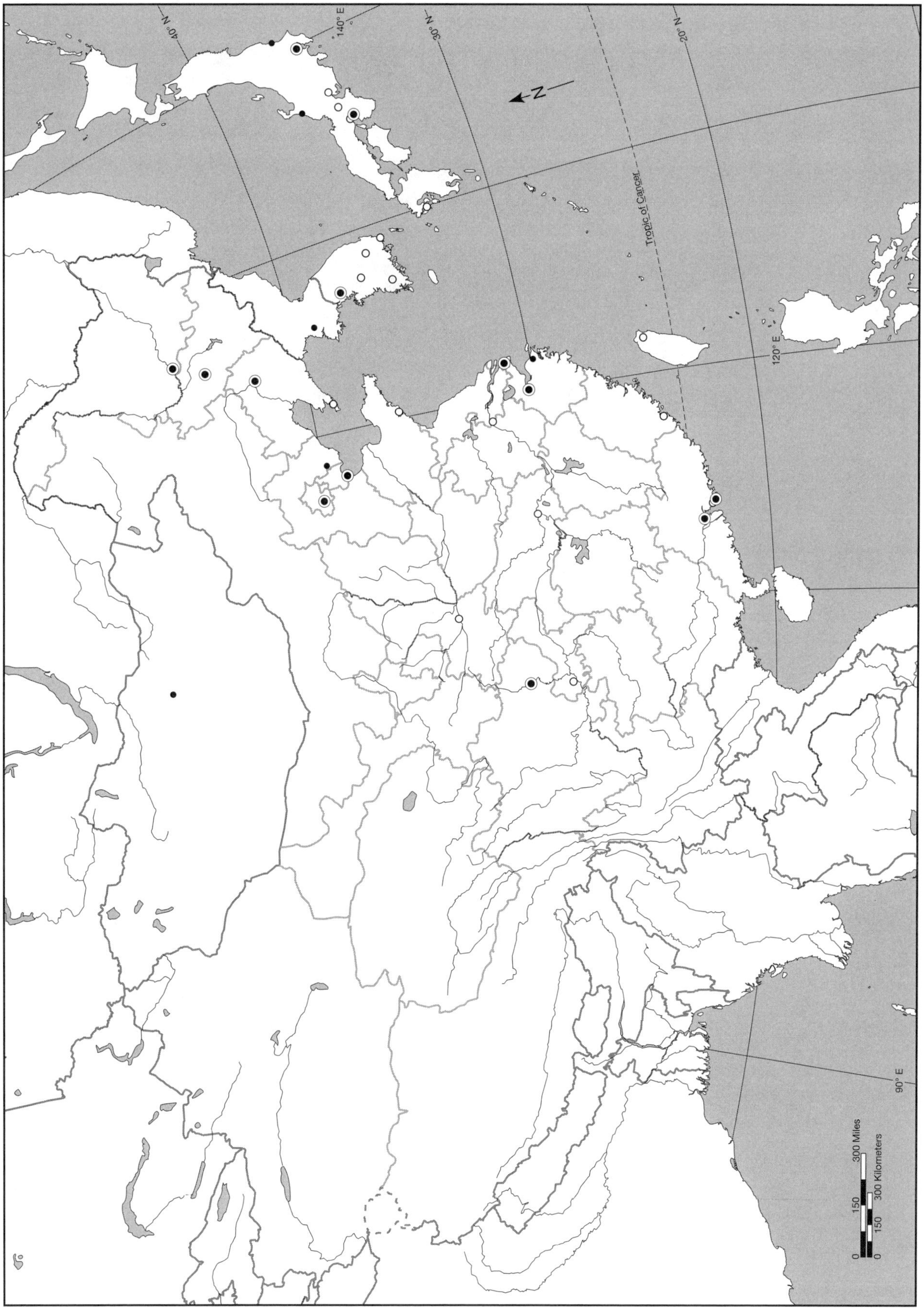

Map 8.1 Countries and Cities

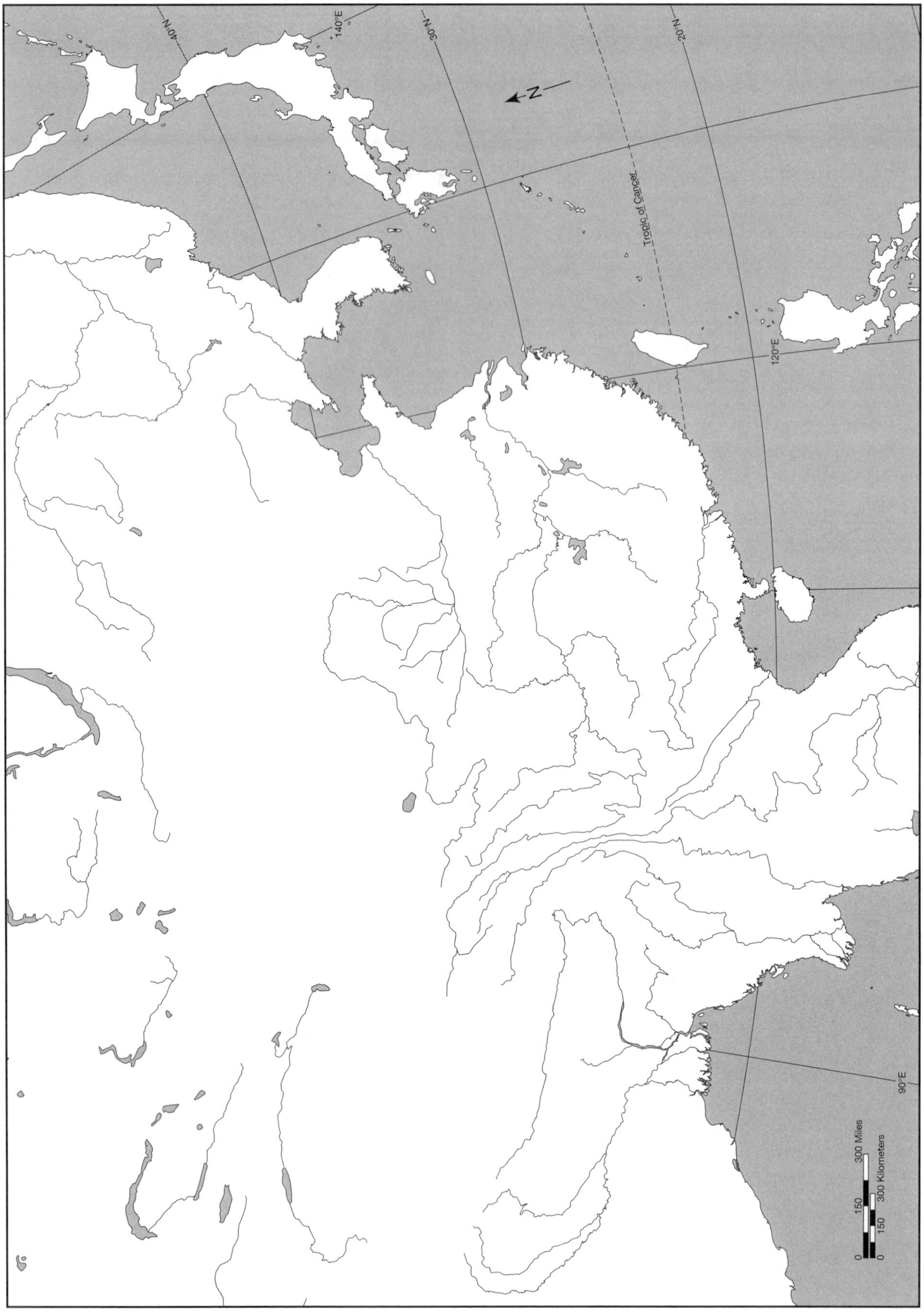

Map 8.2 Landforms and Waterways

Mapping Urbanization

Goal: To spatially locate the region's main urban centers. For this exercise you will need the 2009 World Population Data Sheet located at the back of this book and a set of colored pencils.

1. Begin by turning to Figure 8.1 (p. 272). Using that map, identify on Map 8.3 below all cities in the region with a population over 1 million. Distinguish between cities that are over 5 million and those that are between 1 and 5 million by using different symbols (in a similar way to how the textbook portrays these differences).
2. Now turn to page 291 in your text and look at Figure 8.18. Using that figure, roughly outline and lightly shade all parts of this region that have a population of over 100 people per square kilometer (use a unique color pencil and then make a map legend and label the legend with the color that corresponds to high population density). Where are these cities located in relation to the overall population density of the region? Explain.

____

____

____

____

____

____

____

____

____

____

____

3. Now turn to page 277 in your textbook and look at Figure 8.5. Using that figure as a guide draw the rough outline of the various physiographic regions on your map (using a new colored pencil for each region making sure to add those colors and their explanation to your map's legend). Is there also a relationship between the physiographic regional map and the location of large urban centers? Explain.

____

____

____

____

____

____

____

4. What has facilitated the growth of urban centers in this region historically? Explain.

5. How do these urban centers service their outlying hinterlands (surrounding subregions)? Explain.

6. What role has China's changing policy of economic reform (in the post-1973 period) played in facilitating the growth of urban centers and the changing map of urbanization in China? Explain.

40° N
140° E
30° N
20° N
N
Tropic of Cancer
120° E
90° E
0 150 300 Miles
0 150 300 Kilometers

Map 8.3

CHAPTER 9

South Asia

Learning Objectives

After reading this chapter, you should be able to:

1. have an appreciation of the complex physical geography of this region, which lends itself on the one hand to isolation in the north of the region and connection through the waterways to the south;
2. understand how people have long adapted themselves to the complex climatic geography, including the monsoons, by creating some of the most productive agriculture in the world;
3. comprehend the complexity of demographic growth and change in the region over time;
4. appreciate the diverse socio-economic and cultural differences that distinguish this region, while also understanding how religion and language and cultural practices as well as poverty impact life in this region; and,
5. explore the future of this region's importance in terms of both its geopolitical as well as its economic significance.

Vocabulary Review

1. In 1876, after Queen Victoria was declared Empress of India, the British established the ____________________, which by 1890 extended their control over the entire region, excepting present-day Afghanistan and Nepal.
2. ____________________________ are river branches that flow away from the main stream.
3. The Taj Mahal is an important symbolic feature of the landscape in India today. It is a legacy of the ____________________ Empire, which promoted the spread of Islam in South Asia beginning in the 15th century.
4. Throughout most of India, people practice ____________________, a polytheistic religion that finds its roots in early Dravidian and Indo-Aryan societies.
5. The center of India's massive movie market is the area known as ____________________ in Bombay.
6. ____________________ is a system of kinship groupings, or jati, that are reinforced by language, region, and occupation.
7. In recent years there has been a call by some for a "new" ______________________ in which genetically-modified crops and new agricultural technologies would be used to increase production.
8. In India the political movement called ____________________ calls on India to unite as an exclusively Hindu nation.
9. South Asia's mountain ranges produce a strong ____________________, causing moist air from the sea to lift and condense. This process produces heavy rainfall in different parts of the region in both the summer and winter monsoon seasons.
10. The _____________________ or "beautiful forest" found in West Bengal and Bangladesh is home to the world's largest tidal mangrove forest.
11. In parts of South Asia it is not uncommon to find children who are placed in a situation of ___________________, whereby their labor is exchanged for a loan.
12. An extremely effective method of increasing yields is to carve hilly landscapes into ___________________, which holds the nutrients found in the soil and allows farmers to control the flow of water.
13. In many parts of southern India the birth rates have reached a state of ________________________ or ZPG.
14. The ________________________ founded in Bangladesh is premised on the use of ________________________, which are small loans, to promote grassroots development.

Activities & Problems

1. Using Figure 9.16, answer the following:

   (a) Name all the dependent states of South Asia under British rule.

   ____

   ____

   ____

   ____

   ____

   ____

   (b) Describe the pattern of annexation between 1753 and 1805 and again after 1805. Consider where that annexation took place.

   ____

   ____

   ____

   ____

   ____

   ____

   (c) Based on the map, which two present-day countries remained free from British rule? ______________ and ______________

2. Using Figure 9.27, describe the pattern of major religions in South Asia.

   ____

   ____

   ____

   ____

   ____

   ____

3. Using Table 9.1 (p. 321), answer the following questions.

   (a) In which country is it most likely that a person would not have access to safe water? ______________

   (b) In which country does the adult illiteracy rate exceed 60%? ______________

   (c) In which countries does the percentage of the population without access to sanitation exceed 65%? ______________, ______________, and ______________

4. Using Figure 9.7, answer the following:

(a) Describe the physiographic subregional geography of South Asia.

______

______

______

______

______

______

(b) In which of the following four physiographic subregions would you find:

(i) Thar Desert ______

(ii) Deccan Plateau ______

(iii) Khasi Hills ______

(iv) Western Ghats ______

(v) Balochistan ______

5. (a) What prompted the Soviet Union to invade Afghanistan in 1979?

______

______

______

______

______

(b) In the post-Soviet period, what political problems have made maintaining peace in Afghanistan so difficult?

______

______

______

______

(c) What tensions has the ongoing U.S. presence in Afghanistan caused in Pakistan?

______

______

______

______

Review Exam

1. Compare and contrast the physical features of the plains and the coastal fringe.

2. Describe the monsoon patterns and the impact of the monsoons on the agricultural geography of South Asia.

3. What are the main causes of ethnic tension in Sri Lanka? Explain.

4. Describe the uneven geography of capitalist development in India.

5. Describe the in-migration of the Aryan peoples into South Asia. Where did they settle and what impact did they have on the landscape?

____

____

____

____

____

____

6. Briefly describe some of the major implications of climate change in the region.

____

____

____

____

____

____

7. Which of the following two countries are nuclear powers?
   (a) India and Afghanistan
   (b) India and Sri Lanka
   (c) Pakistan and Afghanistan
   (d) Pakistan and India

8. Which of the following is true of Harappan society?
   (a) It emerged in the Indus River Valley 4500 years ago and eventually opened up trading with Mesopotamia.
   (b) It emerged in the Indus River Valley 4500 years ago and survived even after floods covered its major cities in mud and silt.
   (c) It emerged in the Ganges River Valley 4500 years ago and eventually opened up trading with Mesopotamia.
   (d) It emerged in the Ganges River Valley 4500 years ago and eventually moved the entire civilization to southern India after environmental disasters made it impossible to live in the north.

9. Under the Mughal Empire and Akbar's rule, in India one could find all of the following except ____________________.
   (a) Persian as the official language
   (b) princely kingdoms kept intact
   (c) taxing on non-Muslims
   (d) an increased status of Islam in the region

10. Pakistan, which means "land of the pure," was established as a result of what geopolitical factor?
    (a) the partition of South Asia by the British based on ethnic and religious groupings
    (b) the partition of South Asia by Gandhi based on ethnic and religious groupings
    (c) the partition of South Asia into Islamic and Buddhist regions
    (d) the partition of South Asia into Hindu and Buddhist regions
11. Which of the following languages would a Buddhist most likely speak in Sri Lanka?
    (a) Hindi (b) Tamil (c) Sinhalese (d) Bengali
12. Which of the following is most true about the economy in South Asia?
    (a) Women are more likely to have access to education than men.
    (b) Women are more likely to participate in the informal economy than men.
    (c) Women are more likely to participate in the formal economy than men.
    (d) Women have fewer restrictions on public action than men.
13. When someone has no collateral to back up a loan they might have to enter into what type of economic relationship?
    (a) indentured servitude (b) bonded labor
    (c) free labor (d) slave labor
14. Which of the following is most true of Sri Lanka?
    (a) It was granted independence by the British in 1997 at the same time that they handed Hong Kong over to China.
    (b) It continues to be plagued by ethnic and religious tensions between Tamils and Sinhalese.
    (c) It is home to a large Islamic population.
    (d) Almost 80% of the population does not have access to safe water.
15. Which of the following is true of Bangalore?
    (a) It is one of the smallest cities in South Asia.
    (b) It is considered to be an economic drain on India.
    (c) It is a center for agricultural production.
    (d) It is home to over 500 high-tech companies.

For each of the following statements, write T for "True" or F for "False" on the space provided.

16. It is possible that families in need of cash might sell their child into bonded labor in South Asia.____________________
17. Only 25% of the region's population lives in rural villages today.____________________
18. In recent years, India has developed a large, thriving middle class.____________________
19. Poverty can be found throughout both rural and urban South Asia.____________________
20. The Portuguese successfully help Christianity on the west coast of India beginning in the late 1400s. ____________________
21. Bangladesh was previously known as East Pakistan.____________________
22. Informal economic activities, such as street vending, are rarely seen on the streets of urban South Asia.____________________
23. In rural South Asia schooling is prolific and literacy rates are quite high.____________________

24. Under the Raj system, British rule was extended and plantation agriculture was expanded throughout South Asia. ____________________

25. Tea was first brought to South Asia by the Portuguese from Europe in the 1600s. ____________________

Key Map Items

Identify the following features on Map 9.1 (countries and cities) and Map 9.2 (landforms and waterways). On your political map, Map 9.1, label the major cities by writing the city names next to their respective symbols. When mapping some of the landforms, you may need to create some sort of symbol system that portrays the landform, such as a ^^^^^^^^^^^^ for a mountain range.

Countries	**Major Cities**	**Landforms**	**Waterways**
As listed in Chapter 9 of your textbook (Figure 9.1)	• Bangalore • Chennai • Colombo • Delhi • Dhaka • Hyderabad • Islamabad • Kabul • Karachi • Kathmandu • Lahore • Mumbai • Srinigar • Thimphu	• Aravalli Hills • Deccan Plateau • Eastern Ghats • Himalaya Mtns. • Hindu Kush Mtns. • Karakoram Range Kirthar Range • Sulaiman Range • Thar Desert • Western Ghats	• Arabian Sea • Bay of Bengal • Brahmaputra River • Ganga River • Godavari River • Indian Ocean • Indus River • Krishna River • Narmada River • Yamuna River

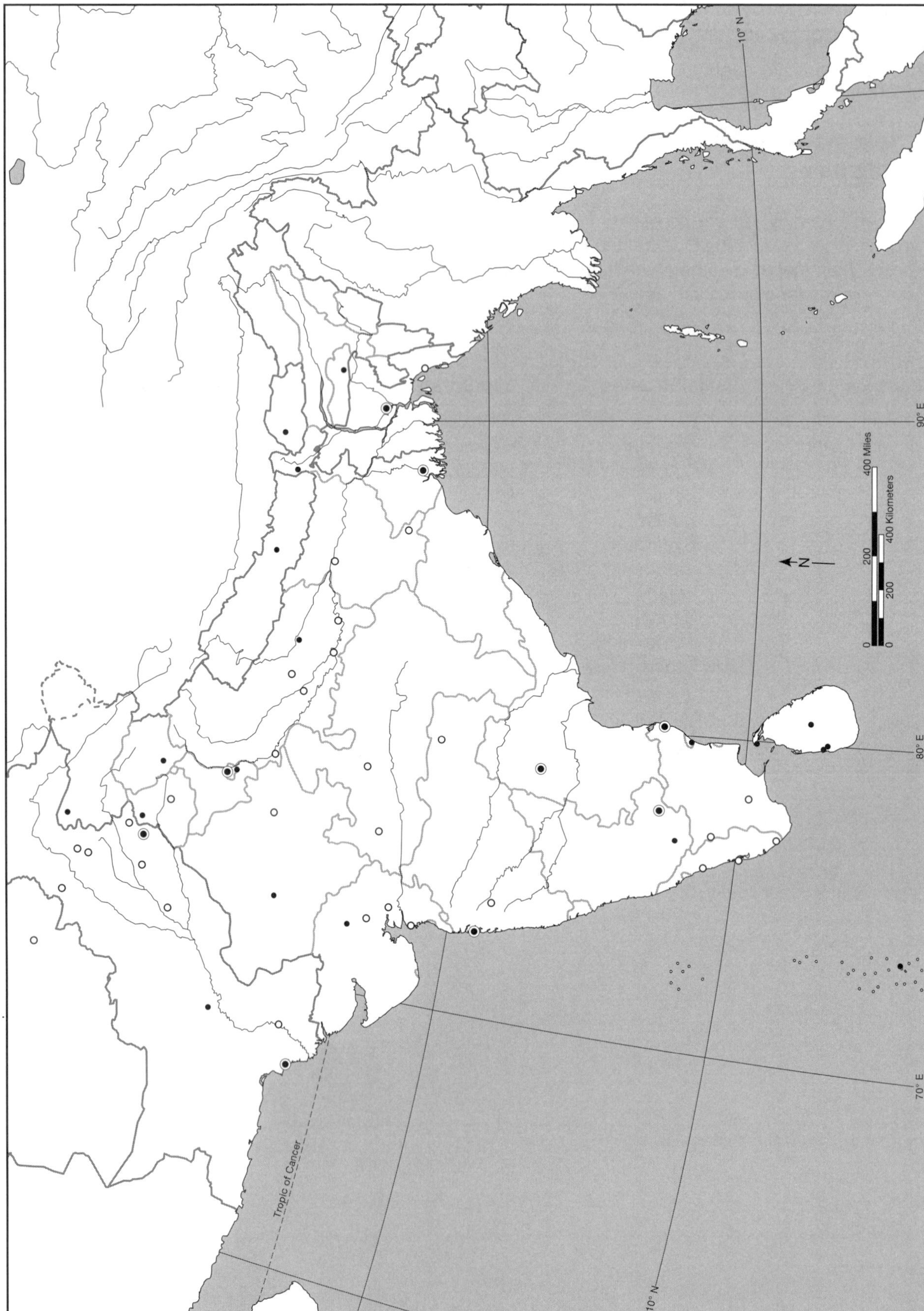

Map 9.1 Countries and Cities

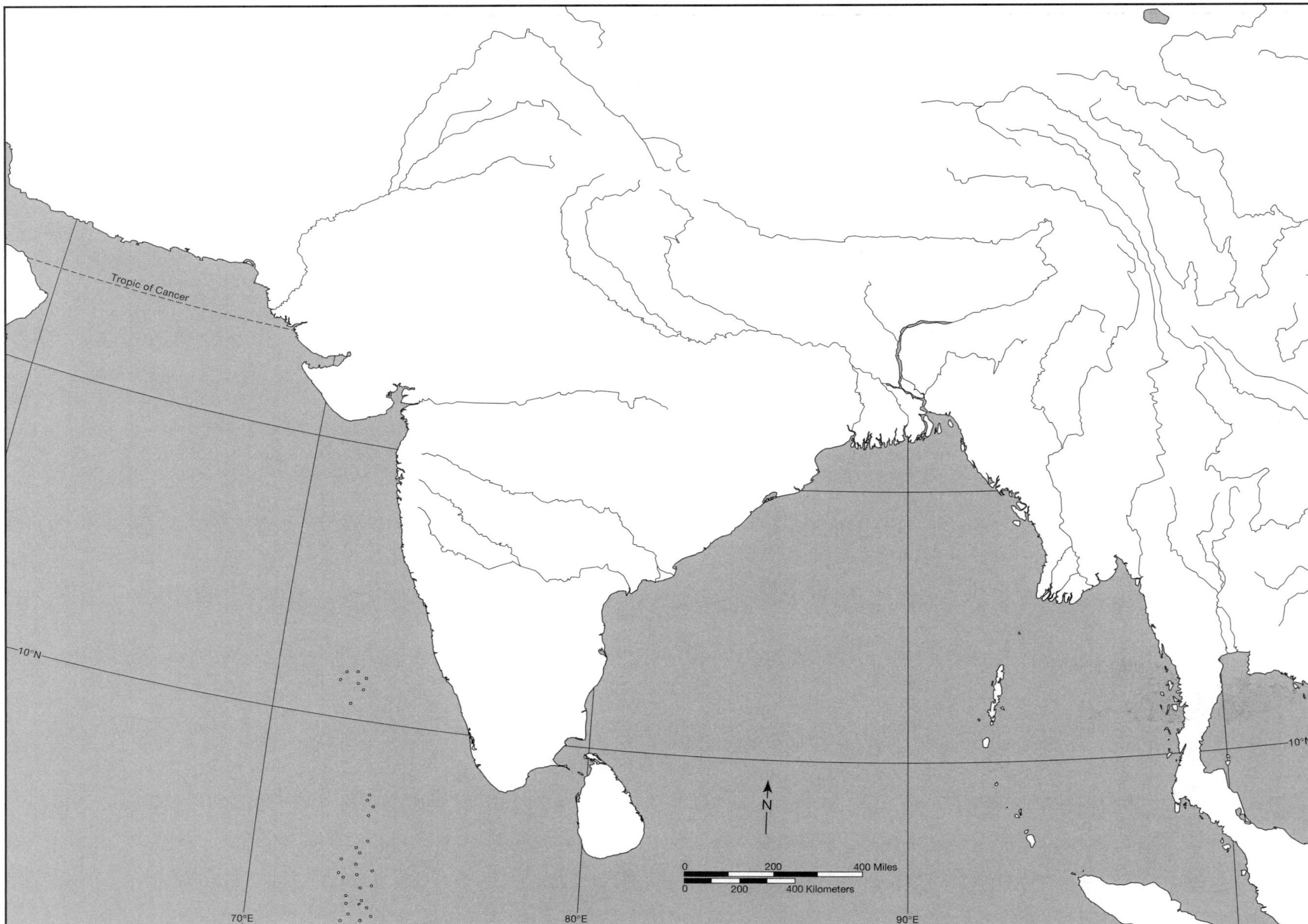

Map 9.2 Landforms and Waterways

Geopolitical Conflict

Goal: In this exercise, you are going to map the significant geopolitical conflict areas in the region and analyze the spatial distribution of those conflicts.

1. Read through Chapter 9 and highlight discussions regarding current geopolitical and or ethnic conflict in the region. Using your imagination, come up with an icon that best represents ethnic conflict in the region. On Map 9.3 place your icons in proximity to the conflict. You may need to create a boundary and/or an explanatory box to highlight the details of the conflict (see example in Chapter 5 on Sub-Saharan Africa, Figure 5.33).

2. What pattern emerges? Are these conflicts concentrated in any particular part of the region?

3. What geopolitical and/or cultural and ethnic factors promote conflict in the region? (e.g., think about borders and cultural distributions). Explain.

4. What impact did European colonialism have on the region? What effect does that colonial legacy have on the map of modern-day conflict you have created? Explain.

5. In what way is the current map of conflict complicated by the modern-day development of nuclear weapons in the region? Explain.

6. What role do international forces, such as the U.S., play in the current regional conflicts? Explain.

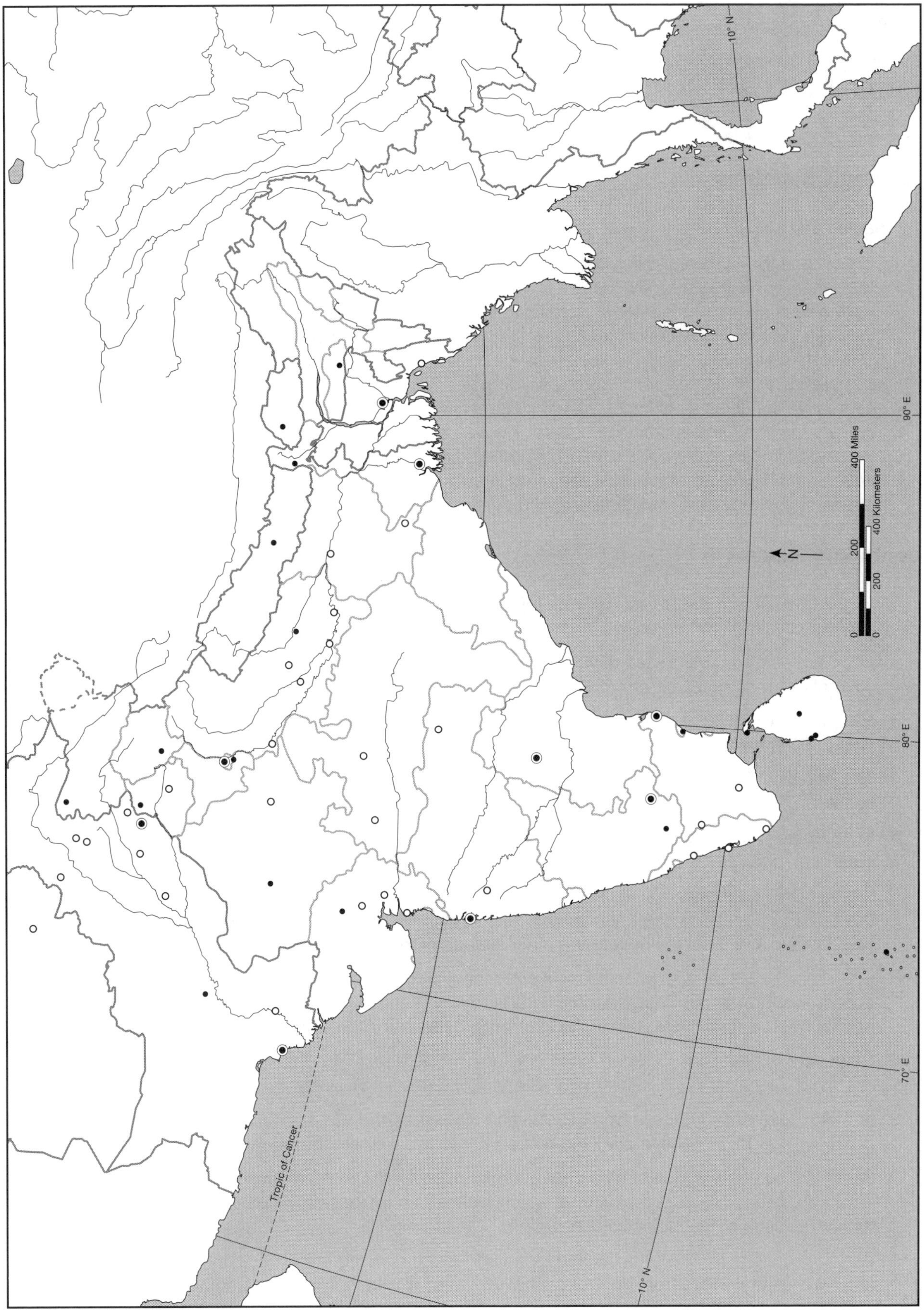
10° N
90° E
80° E
70° E
10° N
Tropic of Cancer
N
0
200
400 Miles
0
200
400 Kilometers

Map 9.3

CHAPTER 10

Southeast Asia

Learning Objectives

After reading and studying this chapter, you should be able to:

1. understand the dynamic and variable physical geography of this region, with particular emphasis on human-environmental relations;
2. have a working knowledge of the historical geography of this region, focusing on the relationship between the region's historical, economic, and demographic development;
3. comprehend the wide range of socio-economic differences in the region, while appreciating the complexity of ethnic, religious, and class identity;
4. appreciate the diversity of linguistic and cultural differences in the region;
5. explore the relationship between cultural practice and social identity, as this relates to health, HIV/AIDS, and minority politics in the region; and,
6. understand the future of regional cooperation and conflict in the region as these relate to broader issues of environmental sustainability and political difference.

Vocabulary Review

1. All the countries of Southeast Asia are influenced by a ____________________ environment, with consistently warm temperatures and seasonal rainfall.
2. ____________________ defines the ecological division between Bali and Lombok, which kept several different species from mixing in the region.
3. Historically, wet rice, or ____________________, has been, and continues to be, the most important crop in Southeast Asia.
4. The islands of Indonesia, which were eventually colonized by the Dutch, were known to Europeans as the ____________________, because of their abundance of nutmeg, cinnamon, and cloves.
5. In 1830 the Dutch instituted the ____________________, whereby all Javanese people must contribute one-fifth of their land and labor to the production of export-oriented commodities.
6. While migration is often a "voluntary" process, in the 1950s the Indonesian government instituted the ____________________, which was designed to redistribute populations out of the densely populated island of Java and into the other islands of the Indonesian archipelago.
7. ____________________ populations were essential to the development of many of the colonial economies throughout Southeast Asia. Later, many of these migrants became part of the entrepreneurial class, establishing banks and insurance companies throughout Southeast Asia.
8. Singapore, a historical entrêpot, is the only city in Southeast Asia that can be considered a ____________________, similar to, perhaps, London and New York City.
9. In 1995, several countries in Southeast Asia established the ____________________ or ___________ to coordinate planning related to the use of their shared river resources.
10. In the last 50 years agriculture has been transformed by the twin processes of land reform and the ____________________, the latter of which focused on increasing yields of staple crops through the introduction of technological innovation.
11. A ____________________ is caused by an underwater earthquake and is defined by a series of large waves that move from deeper to shallow water sometimes making landfall.
12. Today, all 10 countries in Southeast Asia belong to the supranational organization called ____________________, which is most commonly referred to by its acronym, ____________________.

13. The __________________ refers to the United States' belief that if Vietnam were to be occupied by communist forces the rest of Southeast Asia would soon succumb to communism.
14. In Cambodia, during the reign of the Khmer Rouge, people were forced into work camps and millions of people of died resulting in a landscape of ____________________ throughout the country.
15. Unlike many other countries in Southeast Asia, Malaysia in the 1970s established __________________, which encouraged __________________ foreign capital to enter Malaysia to establish various industrial programs.
16. Within the region, ethnic minorities are often euphemstically referred to as __________________ or __________________.
17. __________________, which is also known as slash and burn agriculture, has historically been used to clear tropical rainforest areas for cultivation.
18. The economy of the Philippines is often described as being dominated by __________________, whereby contracts and economic concessions are given to friends and families. Often this leads to the development of a ______________, wherein leaders have diverted national resources to the own personal gain.
19. The rise of HIV rates throughout Southeast Asia has led to a where parents are dying before their children. This results in high number of _______________, who are left to be cared for by grandparents, friends, or the government.
20. Throughout Southeast Asia, ____________________ were brought in to clear forests and help develop the food-centered agricultural potential in what eventually became known as the _______________ of the region.
21. In 2004, a massive _____________ or tidal wave, hit the coasts of Indonesia, Thailand, Burma, Sri Lanka, and India killing tens of thousands of people.
22. During the Cold War, the U.S. became involved in part because the __________________ suggested that if South Vietnam was to fall to the communists, communism would spread further into Southeast Asia and eventually to Australia and the Americas.
23. In Southeast Asia it is common to find people who practice both ____________________ (i.e., a married couple lives with the woman's parents) and ____________________ (i.e., a married couple lives with the man's parents).
24. The Dutch introduced the ________________________ in 1830, whereby all local farmers would have to devote one-fifth of their land and their labor to export crop production.
25. The concept of _________________ has been employed in Malaysia to ensure that the benefits of economic development go first and foremost to those people of Malay ancestry. This practice has intentionally discriminated against Malaysia's Chinese population.

Activities & Problems

1. Using Figure 10.3, answer the following questions.
   (a) What type of climate would you find in Vietiang? __________________
   (b) What type of climate would you find in Jakarta? __________________
   (c) What type of climate would you find in Rangoon? __________________
   (d) What type of climate would you find in Hanoi? __________________
   (e) On what islands would you find highland climate regions? __________________
2. Using Figure 10.10, describe the vegetation geography of Southeast Asia. What impact have humans had on the overall ecology of the region?

____

3. Using Figure 10.11, answer the following:
   (a) Which country has the most forest cover in total hectares? ____________
   (b) Which country has the least forest cover in total hectares? ____________
   (c) Which country has the highest deforestation rate between 1973 and 1985? ____________
   (d) Which country has the lowest deforestation rate between 1973 and 1985? ____________
   (e) Which country has the highest deforestation rate between 1990 and 2000? ____________
   (f) Which country has had an increase in forest cover between 2000 and 2005? ____________
   (g) Which countries have had less than a 5% deforestation rate between 2000 and 2005? ______
4. Using Figure 10.13, name the countries that were colonized by the following European powers.
   (a) French ____________
   (b) British ____________
   (c) Dutch ____________
   (d) Spanish ____________
   (e) United States ____________
   (f) According to the map, in what year were the Philippine Islands first colonized? ____________ In what year did the colonial power in the Philippines change hands? ____________
5. Using Figure 10.14, describe the pattern, scope, and extent of Japanese occupation of Southeast Asia during WW II.

Review Exam

1. Describe the different ways in which monsoons impact the mainland and insular regions of Southeast Asia.

2. Describe the relationship between population density and deforestation in Southeast Asia. What areas have witnessed the highest levels of deforestation in the region? Which regions have witnessed the lowest levels?

3. Describe Indonesia's transmigration program. What are some of the problems with the program?

4. What role has overseas Chinese played in the political and economic development of Southeast Asia?

5. What world regions have received the largest number of labor migrants from Southeast Asia? What kinds of jobs are these people doing? How is the work gendered?

6. Southeast Asia is ethnically diverse and many countries have minority ethnic populations within their borders. What are some of the issues related to minority politics in Southeast Asia today?

7. Discuss the problem of HIV in Southeast Asia. How does this social problem relate to the sexual politics and drug use politics in the region?

8. Which of the following is not a conflict zone in Southeast Asia?
   (a) the Muslim rebellion in Mindanao, Philippines
   (b) the Shan rebellion in northern Burma
   (c) the Thai Muslim rebellion near the Laos border in northeastern Thailand
   (d) the Karen rebellion along the Thai/Burma border
9. Which three Southeast Asian countries constitute the "Golden Triangle"?
   (a) Burma, Thailand, and Laos
   (b) Burma, Thailand, and Malaysia
   (c) Thailand, Laos, and Vietnam
   (d) Thailand, Laos, and Cambodia
10. Which of the following statements best reflect the physical geography of Southeast Asia?
   (a) Burma and Thailand are part of the Ring of Fire.

(b) The Philippines is in a major typhoon track.

(c) Burma is subject to extreme earthquakes.

(d) Most of Laos is at sea level.

11. Which of the following best describes the culture system?

    (a) This system required all Javanese to learn Dutch.

    (b) This system required all Javanese to wear Western dress.

    (c) This system required all Javanese to promote Dutch culture.

    (d) This system required all Javanese to devote one-fifth of their land and labor to export crop production.

12. Which of the following best describes what happened during the rule of the Khmer Rouge in Cambodia between 1975 and 1979?

    (a) Only a few people lost their homes and jobs and most people adapted to the new economic conditions.

    (b) Tens of thousands of refugees migrated out of Cambodia and into Burma.

    (c) Over 2 million people were killed during the period, when cities were cleared and many people were put to work on rural communes.

    (d) (a), (b), and (c).

13. Overseas Chinese communities are best described by which of the following statements?

    (a) In most cases, Chinese have assimilated into all host societies with little problem.

    (b) Many ethnic Chinese in SEA were important actors in the colonial economies of the region.

    (c) Singapore and Malaysia became one country in 1982 to unite ethnic Chinese in Singapore with ethnic Malays in Malaysia under one common political and economic system.

    (d) Ethnic Chinese make up only a small percentage of the population of the island of Singapore today.

14. Which of the following is the most accurate statement about Southeast Asia's overall economic level?

    (a) Southeast Asia boasts some of the highest and some of the lowest gross national products in the world.

    (b) Overall the economies of Southeast Asia are all relatively equivalent.

    (c) The former socialist countries have enjoyed the most rapid economic growth since 1975 in the region.

    (d) Ethnic minorities in Southeast Asia, such as hill tribes in the mainland countries, are more likely to enjoy economic success than the majority populations.

15. Which of the following best represents the Mekong basin?

    (a) Hanoi is served by this large river.

    (b) The headwaters begin in China and end in the South China Sea.

    (c) There has been little dialogue between the countries in the basin region.

    (d) Since the river is slow there is little potential for hydroelectric power.

16. A person living in northern Vietnam would most likely practice which religion?

    (a) Hinduism
    (b) Therevada Buddhism
    (c) Islam
    (d) Mahayana Buddhism

17. Which of the following best reflects the consequence of Spanish rule in the Philippines?

    (a) They spread Catholicism and gained access to spice trade routes.

    (b) They spread Catholicism and stopped the flow of spices out of the region.

(c) They defended local religious practices and facilitated the growth of indigenous political systems in the region.

(d) They defended local religious practices and expanded the Spanish Empire.

For each of the following statements, write T for “True” or F for “False” on the space provided.

18. The British colonized Thailand in the 19th century. ____________________
19. The concept deva-raja, or god-king, was brought to Southeast Asia from India. ____________________
20. Someone living in Java would most likely practice Islam. ____________________
21. Burmese is an Austro-Asiatic language. ____________________
22. Japan is often blamed for deforestation in Southeast Asia because of its high demand for hardwoods. ____________________
23. Bali is the only place in Southeast Asia where Hinduism is practiced. ____________________
24. In most cases, Southeast Asia is marked only by matrilocality, where married couples live with the woman's family.____________________
25. Only about 5% of the population of Southeast Asia currently lives in an urban area. ____________________
26. Thailand has been a net recipient of refugees in the last 20 years. ____________________
27. Under British control, Malaysia largely produced rubber and tin. ____________________
28. Between 1990 and 1992 Cambodia experienced an annual percent change in inflation that soared over 150%. ____________________
29. Burma's stable government means that there are few people who flee that country and become refugees in other countries. ____________________
30. Urban growth in the region will have no impact on the future of environmental sustainability in the region. ____________________

Key Map Items

Identify the following features on Map 10.1 (countries and cities) and Map 10.2 (landforms and waterways). On your political map, Map 10.1, label the major cities by writing the city names next to their respective symbols. When mapping some of the landforms, you may need to create some sort of symbol system that portrays the landform, such as a ^^^^^^^^^^^^ for a mountain range.

Countries	**Major Cities**	**Landforms**	**Waterways**
As listed in Chapter 10 of your textbook (Figure 10.1)	• Bandar Seri Begawan • Bangkok • Cebu City • Chiang Mai • Hanoi • Ho Chi Minh City • Hue • Jakarta • Kuala Lumpur • Luang Prabang • Mandalay • Manila • Penang • Phnom Penh • Rangoon • Singapore • Vientiane	• Arakan Mountains • Annam Mountains • Bali • Barisan Mountains • Borneo • Isthmus of Kra • Java • Mindanao • Moluccas • Timor • Sulawesi • Sumatra	• Andaman Sea • Banda Sea • Celebes Sea • Chao Phraya River • Gulf of Thailand • Indian Ocean • Irrawaddy River • Java Sea • Mekong River • Red River • South China Sea • Straits of Malacca • Sulu Sea • Tonle Sap Lake

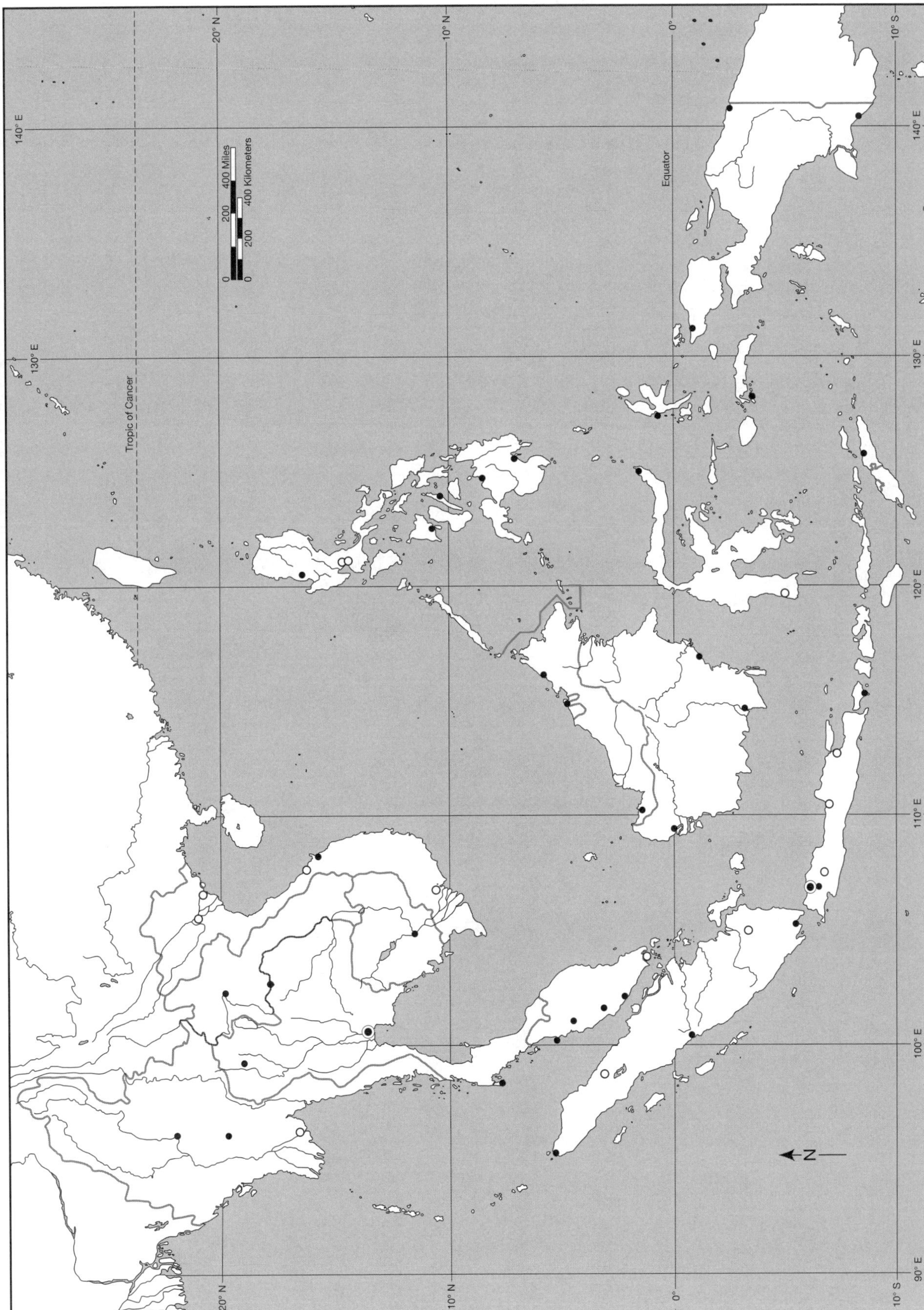

Map 10.1 Countries and Cities

Map 10.2 Landforms and Waterways

Mapping Urbanization

Goal: Urbanization across Southeast Asia is dramatically different. In this exercise, you are going to map the "% Urban," using the 2009 World Population Data Sheet at the back of this workbook.

1. Begin by turning to the 2009 World Population Data Sheet and locate the region "Southeast Asia." Go to the column "% Urban," and copy down the rates of urbanization for each country in the region in rank order. Create three categories of data for "highly urbanized," "moderately urbanized," and "marginally urbanized."
2. Create a key for each one of the data categories and shade that category accordingly (i.e., the lightest shade should be used for the smallest category). A key should also be placed on your map (see discussion above and samples in the text of choropleth mapping in the appendix of your textbook).
3. Using Map 10.3, create a map of your data by labeling the countries according to your categorization. Also, give your map a title.
4. Describe the pattern of urbanization. What limitations are there to using a national level map for analyzing urbanization patterns? Explain.

5. What accounts for the different rates of urbanization in the region? Explain with examples from the text that support your argument.

6. Using Figure 10.1, identify all cities on Map 10.3 with a population of 1 million or more people. Is there a relationship between these large urban centers and the map of overall urban density? Explain.

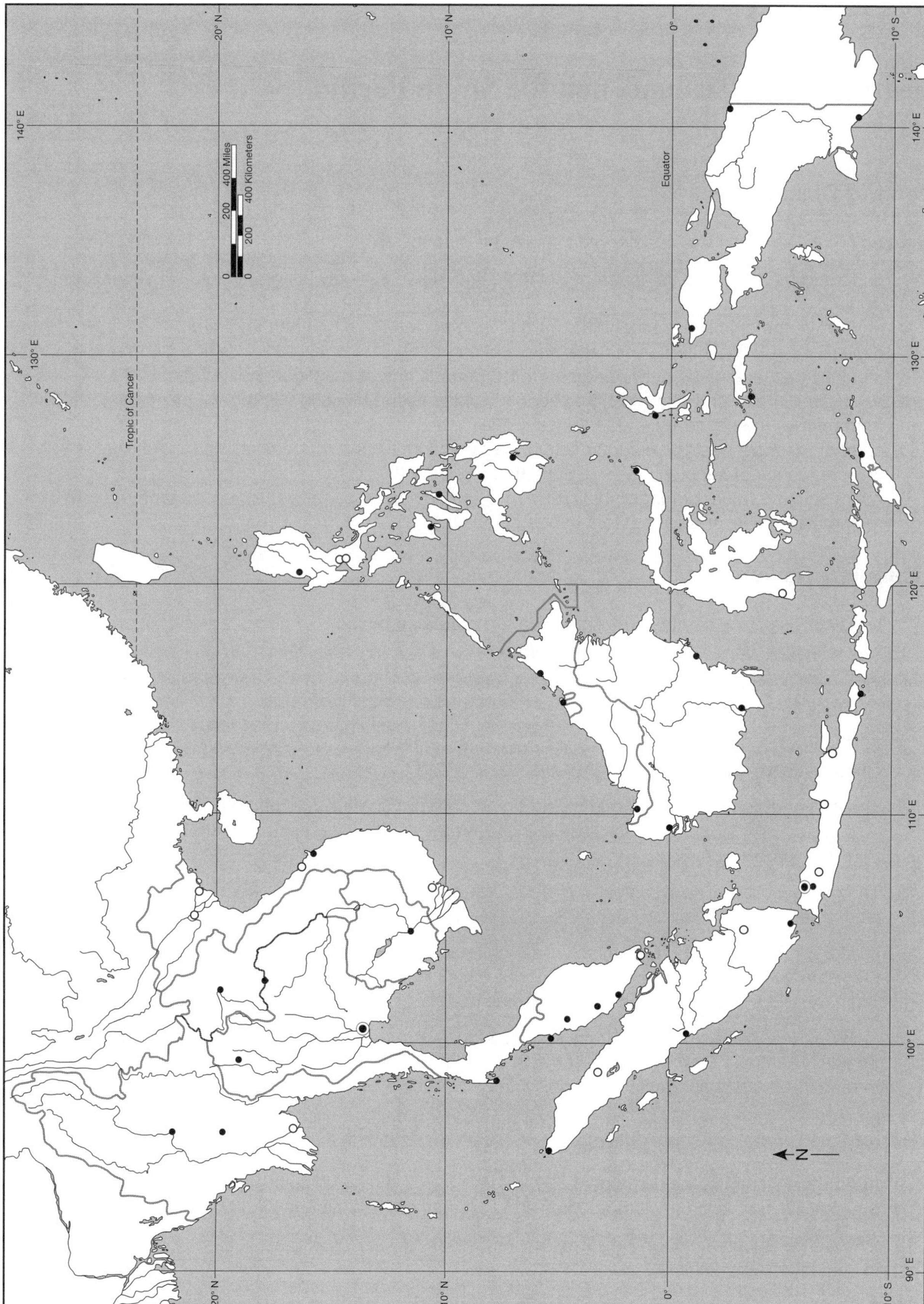

20° N
10° N
0°
10° S
140° E
130° E
120° E
110° E
100° E
90° E
Tropic of Cancer
Equator
0 200 400 Miles
0 200 400 Kilometers
N

Map 10.3

CHAPTER 11

Australia, New Zealand, and the South Pacific

Learning Objectives

After reading and studying this chapter, you should be able to:

1. understand that despite its apparent isolation, this region is highly integrated into the wider Earth system;
2. appreciate the significant impact that introduced species have had on the region;
3. explore the relationship between the historical migration of peoples to this region and the ongoing tensions that take place between different indigenous and non-indigenous groups;
4. have a better comprehension of the economic shifts taking place in this region, as economies shift from primary to secondary and tertiary activities;
5. compare and contrast the differing sets of social and cultural practices that distinguish the very diverse groups of peoples in this region; and,
6. appreciate the future demographic challenges this region faces as well as understand the role this region plays in wider development efforts globally.

Vocabulary Review

1. Land claims on the continent of Antarctica are regulated by the ____________________, which was signed in 1958 and includes 44 countries.
2. Throughout the outback of Australia one can find ____________________, where ranchers raise livestock on vast open ranges.
3. In Oceania most of the fishery resources are managed as ____________________, which affords all community members equal rights to access.
4. The entire region is vulnerable to global environmental changes. In particular, ____________________, which results in higher levels of ultraviolet radiation, has become a particular problem in Oceania.
5. The group of Aboriginal peoples who were forced into foster homes between 1928 and 1964 is often referred to as the ____________________.
6. Unlike the South Pacific Commission, which included the United States, France, and the United Kingdom, the ____________________, which was established in 1971 as an organization to promote regional cooperation, does not include any of the former colonial powers in the region.
7. In 1983, New Zealand and Australia built upon their supranational economic organization, the New Zealand–Australia Free Trade Agreement, by signing the ____________________, which set out to remove all tariffs and restrictions on trade between the two countries.
8. In the early 1900s, the Australian government set forth the ____________________, which was intended to maintain Australia's European "look."
9. In Aboriginal culture, the concept of ____________________ joins the past and future, people and places, in a continuity that ensures a high respect for the natural world.
10. In 1840 the British annexed New Zealand as part of the ____________________, which was signed by 40 Maori chiefs.
11. A(n) ____________________ is defined as an island that is made up of a coral reef.
12. Under European rule many new species were introduced to the region in a process some call ____________________. This process led to the endangerment or extinction of numerous indigenous species in the region.

13. One of the most unique features of Australia is the proliferation of ____________________, animals that give birth to premature offspring that must then be fed by nipples located in an outer pouch on the mother's body.

14. Based on the theory of ____________________, it is believed that larger islands will have a greater degree of biodiversity than smaller islands.

15. On the island of Nauru, deep deposits of ____________________ or bird droppings have resulted in the highest-quality phosphate rock in the world.

16. ____________________ refers to the group of islands north of the equator.

17. Papua New Guinea and Fiji are just two islands that are part of ____________________, a region of islands found in the western Pacific.

18. The area in Australia known as the ____________________ is so named because it is underlain by the world's largest aquifer.

19. Even though income levels are low throughout much of the Pacific islands, the area is considered to be a place of ____________________ because people can rely on the rich supply of fish and coconuts to sustain a decent diet and support relatively high life expectancies.

20. Although most ______________________ sites are of historical and cultural significance, the Great Barrier Reef is also protected by the United Nations.

Activities & Problems

1. Using Figure 11.3, answer the following questions.
   (a) What type of climate and vegetation would you find in Cairns? ____________________ and ____________________
   (b) What type of climate and vegetation would you find in Alice Springs? ____________________ and ____________________
   (c) What type of climate and vegetation would you find in Broome? ____________________ and ____________________
   (d) What type of climate and vegetation would you find in Sydney? ____________________ and ____________________
   (e) What type of climate and vegetation would you find in Brisbane? ____________________ and ____________________
   (f) What type of climate and vegetation would you find in Hobart? ____________________ and ____________________

2. Using Figure 11.5, answer the following:
   (a) Describe the physical landscape geography of Australia.

____

____

____

____

____

____

(b) Describe the physical landscape geography of New Zealand.

____

____

____

____

____

(c) Using the text and Figures 11.3 and 11.5, compare and contrast the physical landscape geographies of Australia and New Zealand.

____

____

____

____

____

____

(d) Which cities in Australia have the greatest populations? Where are they located? Why do you think those places have fostered such large populations relative to the rest of the country?

____

____

____

____

____

____

3. Using Figure 11.14, answer the following questions.

(a) In what year did Cook make it to the eastern coast of Australia? ______________

(b) Where did Cook's "third voyage" end? ______________

(c) In what year did the British claim (and settle) Sydney? ______________ Newcastle? ______________ Hobart? ______________ Darwin? ______________ Perth? ______________

(d) In what year did the Dutch claim Van Diemen's Land? ______________ How far east did Abel Tasmen get on his voyage? ______________

4. Using the subsection "Geographies of Indulgence, Desire, and Addiction: Uranium in Oceania," pages 378–379, answer the following questions:

   (a) What is uranium and why did it become so important in the post-WW II period? Where can one find uranium today in this region?

   (b) What is significant about nuclear testing in this region?

   (c) What have been some of the consequences of nuclear testing in the region?

   (d) What have some private organizations and local governments done to limit nuclear testing in the region?

____

____

____

____

(e) What have been the results of the protests against nuclear testing in the region?

____

____

____

____

____

____

5. Based on your reading of the Signature Region in Global Context box on pps. 396–397, answer the following questions.
   (a) What is the average temperature during the six-month winter? _______________
   (b) Which country has a base at the Amundse-Scott South Pole _______________ and the Geomagnetic South Pole _______________
   (c) About how many tourists visit Antarctica a year? _______________
   (d) Which South American country has a claim in Antarctica? _______________
   (e) Which bird is considered the "indicator species" for measuring human disturbance on Antarctica? _______________

Review Exam

1. What are some of the main issues related to indigenous rights in Australia and New Zealand?

____

____

____

____

____

2. Briefly examine the various economic practices associated with early colonialism in Australia and New Zealand and the implications for in-migration to the region.

____

____

____

3. What were the implications of the Treaty of Waitangi in New Zealand?

4. How have the Pacific Islands been integrated into the world economy?

5. How were Australia's immigration policies modified in 1973? What was the impact of that change?

6. What are some of the future issues that face this region? What role might leaders in this region play globally in the future?

7. In what ways has the restructuring of global agriculture trade impacted New Zealand?
   (a) There has been an increase in the export of traditional exports such as lamb meat.
   (b) There has been an increase in the export of traditional exports such as wool.
   (c) There has been an increase in the export of new exports such as wine and kiwi.
   (d) There has been an increase in the export of new exports such as wool and wine.
8. Which of the following is true of ecological imperialism in the region?
   (a) The dingo was first introduced by Europeans in the 19th century.
   (b) The region's flightless birds were most vulnerable to the influx of rats, cats, dogs, and snakes.
   (c) During the colonial period Europeans limited the introduction of weeds, pests, and non-indigenous crops.
   (d) The introduction of the rabbit to Australia has helped aid in reducing other nonindigenous species.
9. Which Australian policy restricted immigration to people from Northern Europe?
   (a) White Australia Policy
   (b) Treaty of Waitangi Policy
   (c) Closer Economic Relations Policy
   (d) Assimilation Policy
10. Which of the following was not a reason that decolonization of the Pacific Islands was slow to occur after World War II?
   (a) The islands were of strategic significance to the United States and other allied nations.
   (b) The islands were of economic significance to the colonial powers.
   (c) The islands were of importance to the advancement of nuclear technology.
   (d) The islands were important sources of migrant labor to the newly booming United States economy.
11. Maori protests broke out in which city in 2009?
   (a) Aukland
   (b) Dunedin
   (c) Christchurch
   (d) Wellington
12. Which of the following is not true of Indo-Fijian and ethnic Fijian relations in Fiji?
   (a) The conflict is a legacy of British colonial rule.
   (b) By 1960 Indo-Fijians almost outnumbered ethnic Fijians.
   (c) As of 2006, a coalition of Indo-Fijians has been running the country.
   (d) There have been 2 coups in the country, in 1987 and 2000.
13. Which of the following countries does not lay claim to Antarctica?
   (a) United Kingdom
   (b) Argentina
   (c) France
   (d) Germany
14. Which of the following is one of the fastest growing industries in the Pacific Island region today?
   (a) tourism
   (b) car manufacturing
   (c) textiles
   (d) mining
15. Which of the following countries does not maintain any colonial or political ties with island nations in the Pacific?
   (a) United States
   (b) Japan
   (c) United Kingdom
   (d) France

16. Which of the following is true of Cook's journey to the Pacific?
    (a) He made three voyages to the area.
    (b) He was killed during a mutiny in Tonga.
    (c) He helped found Hobart.
    (d) He took prisoners and established Sydney in 1788.

For each of the following statements, write T for "True" or F for "False" on the space provided.

17. New Zealand is less than 50% urban. ____________________
18. A marsupial gives birth to a premature offspring. ____________________
19. The theory of island biogeography states that larger islands are likely to have higher levels of biodiversity. ____________________
20. The demand for Australian wool was driven by the industrial revolution in the United States during the 19th century. ____________________
21. In the 1970s the economies of the region began to shift from Europe to Asia.
    ____
22. The Maori in New Zealand have much less power and recognition then Aborigines in Australia.
    ____
23. Southeastern Queensland is sometimes referred to as the "Gold Coast." ____________________
24. Of all the countries in the world, Malaysia sent the most tourists to Australia in 2007.
    ____
25. In the 1920s both Australia and New Zealand made conscious efforts to develop local industry through the development of import substitution policies. ____________________

Key Map Items

Identify the following features on Map 11.1 (countries and cities) and Map 11.2 (landforms and waterways). On your political map, Map 11.1, label the major cities by writing the city names next to their respective symbols. When mapping some of the landforms, you may need to create some sort of symbol system that portrays the landform, such as a ^^^^^^^^^^^^ for a mountain range.

Countries	**Major Cities**	**Landforms**	**Waterways**
As listed in Chapter 11 of your textbook (Figure11.1)	• Adelaide • Alice Springs • Auckland • Brisbane • Cairns • Canberra • Darwin • Melbourne • Papeete • Perth • Port Moresby • Sydney • Wellington	• Ayres Rock • Blue Mountains • Eastern Highlands • Great Barrier Reef • Great Dividing Range • Great Victorian Desert • Interior Lowlands • MacDonnell Ranges • Melanesia • Micronesia • Polynesia • Southern Alps • Tasmania	• Arafura Sea • Bass Strait • Cook Strait • Coral Sea • Great Australian Bight • Indian Ocean • Timor Sea • Timor Sea • Tasman Sea

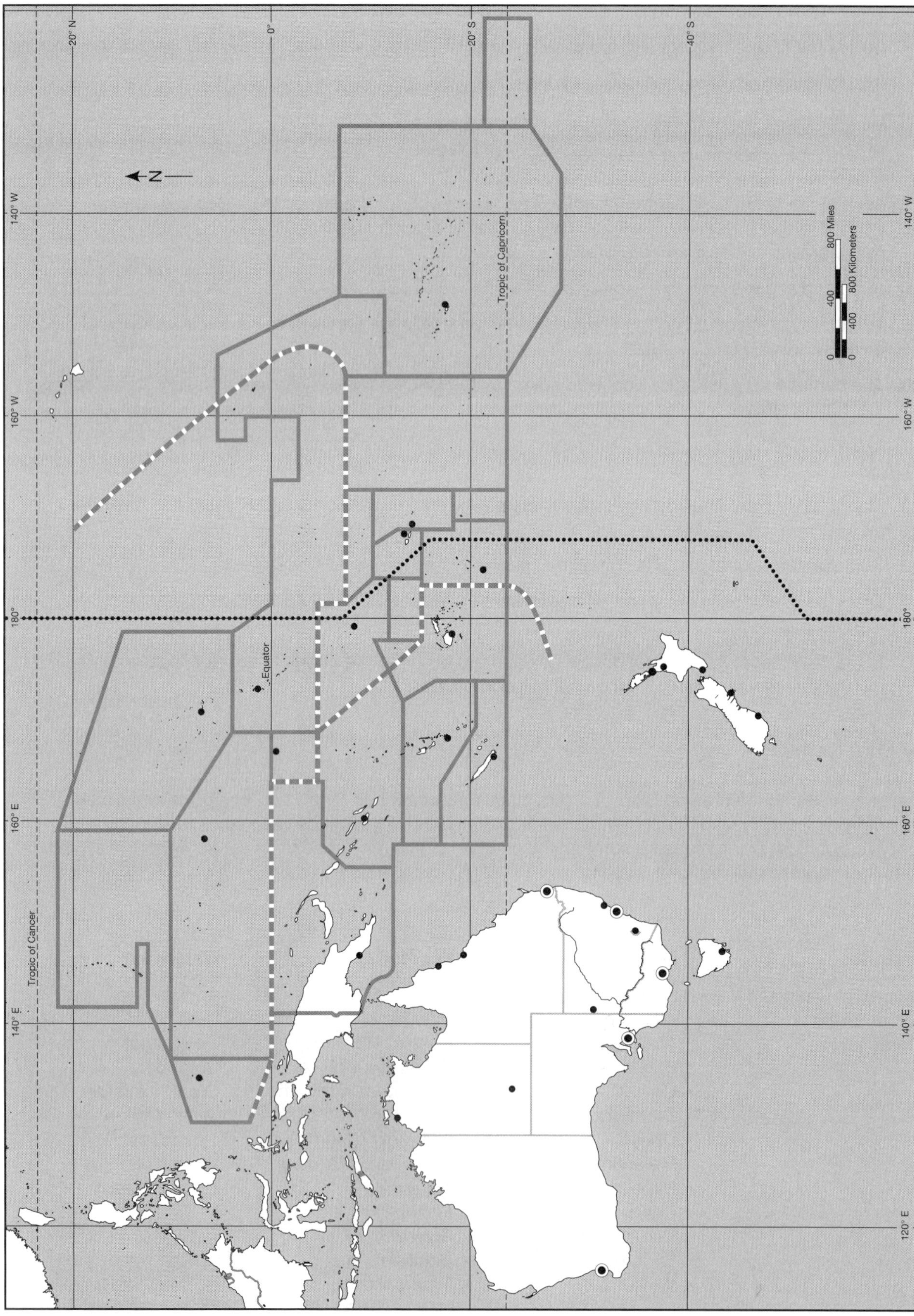

Map 11.1 Countries and Cities

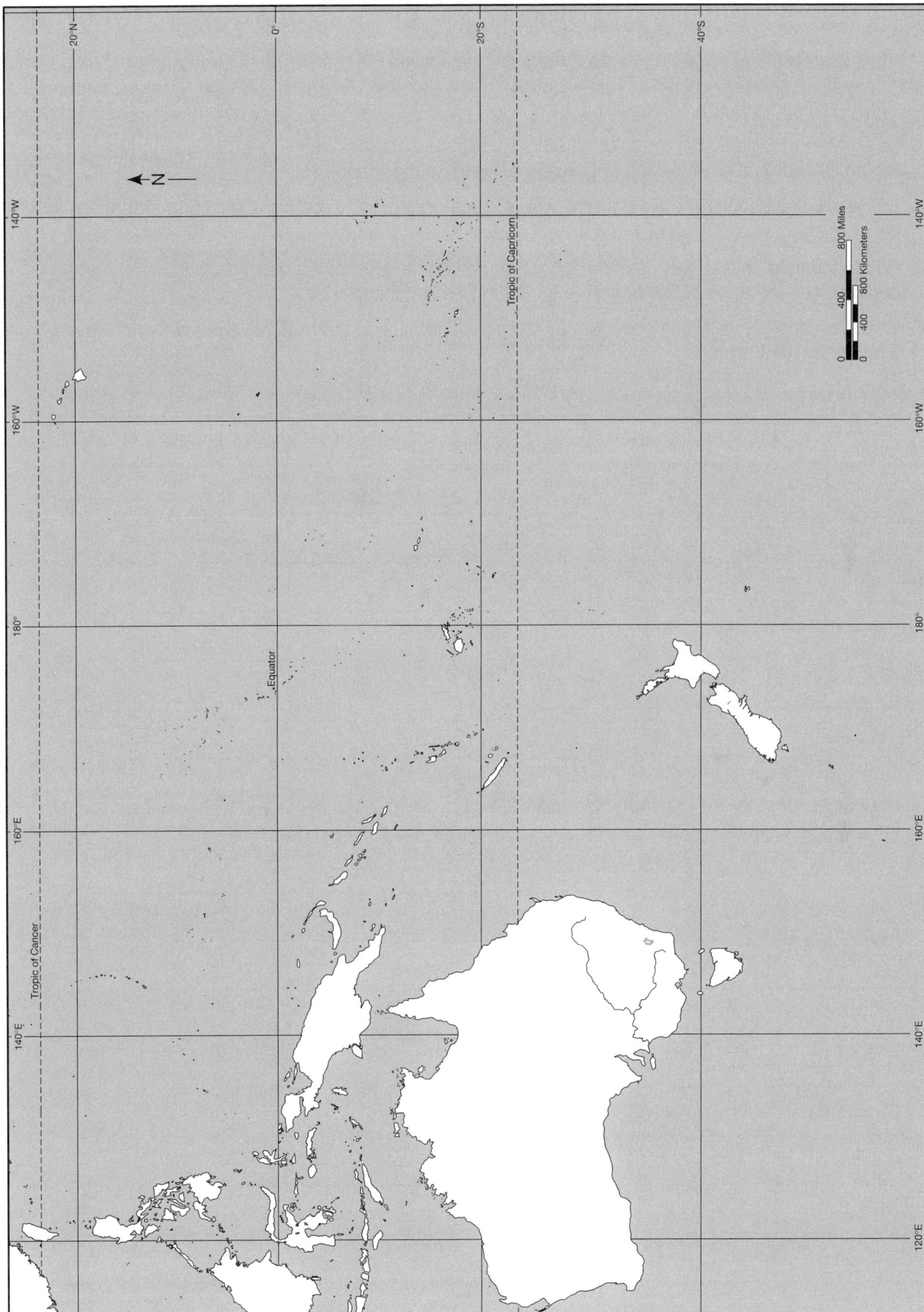

Map 11.2 Landforms and Waterways

Mapping Human-Environment Relations

Goal: In this exercise, you are going to map vegetation geography in relation to the geography of primary commodity production in Australia. You will want to do a detailed reading of Chapter 11, particularly Figures 11.3, 11.5, 11.15, and 11.25.

1. Using Map 11.3 and different colored pencils outline the climatic and vegetation geography of the region, providing a key for each region represented on the map.
2. Using points and outlines, highlight the wheat, sugarcane, sheep, cattle, and mining production in the country.
3. Add to the map, in text, key physiographic features that might have a direct impact on the primary commodity geography of this region (e.g., Great Dividing Range).
4. Where are the commodities you mapped developed relative to the map of vegetation and climate? Explain the relationships.

5. How have humans had to adapt their environment to develop the primary industries discussed in Map 11.15? Explain.

6. What international factors affected the spatial organization of primary commodity production in Australia? Explain.

7. Where are aboriginal lands located in relation to the various vegetation, climatic, and resource geographies you have already mapped? Explain.

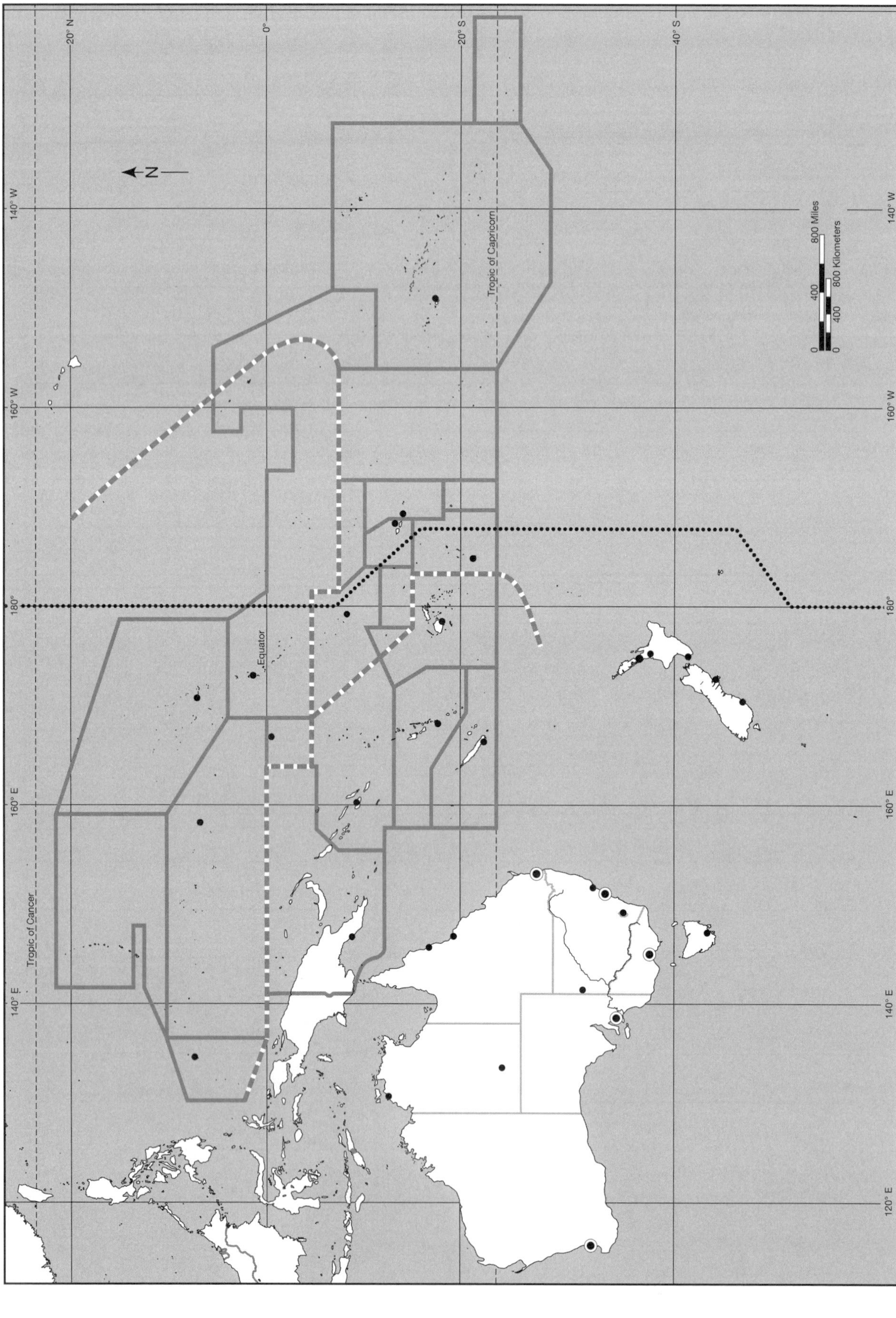
N
20° N
0°
20° S
40° S
140° W
160° W
180°
160° E
140° E
120° E
Tropic of Cancer
Equator
Tropic of Capricorn
0 400 800 Miles
0 400 800 Kilometers

Map 11.3

Appendix A: Answers

Preface

Answers to "Testing Your Map Knowledge"

1. latitude; longitude
2. prime meridian; equator
3. nominal; absolute
4. global positioning system
5. site and situation
6. physical attributes; a place relative to other places and human activity
7. cognitive maps
8. equal-area; outline
9. Remote sensing
10. c
11. (i) a, (ii) d
12. data capture
13. geodemographic research

Chapter 1: World Regions in Global Context[1]

Vocabulary Review

1. region; places
2. two superpowers, the U.S. and U.S.S.R., built blocs and alliances around their own ideological differences
3. regionalism
4. periods of international power established by individual states through economic, political, and military competition
5. primary; secondary
6. demographic transition
7. GDP; GNP
8. irredentism
9. commodity chain
10. global south
11. plate tectonics
12. weather; climate
13. intertropical convergence zone; ITCZ
14. biome; biogeography
15. global warming; greenhouse effect
16. world-system
17. cultural hearth areas

[1]As noted above, answers to the Key Map Items and various Mapping Exercises that follow the Review Exams of each chapter are not included in the answer key. The responses to these questions will vary based on your own analysis.

18. urbanization; colonization
19. diaspora; nationalism
20. Hegemony
21. division of labor
22. neocolonialism
23. Before Common Era; Classical
24. World Bank; International Monetary Fund; IMF
25. Neoliberalism
26. transnational corporation

Activities & Problems

1. North America, most of Western Europe, East Asia (Japan and South Korea). Australia and New Zealand.
2. Answers will vary.
3. (a) divergent
   (b) subduction
   (c) subduction
   (d) subduction
   (e) subduction
4. See textbook, page 4, for answers.
5. As sinking air moves from high to low pressure cells some of that air is forced back toward the equator, forcing it to rise once again, creating a band of constant high pressure around the equator
6. Any three of the following: Nigeria, Democratic Republic of Congo, Central African Republic, Cameroon, Gabon, Republic of Congo, Equitorial Guinea
7. Any three of the following: Cambodia, Burma, Laos, Vietnam, or Thailand
8. French in 1714: Europe; French in 1914: Western/Northern Africa, Madagascar, Southeast Asia, South America; Spanish in 1714: North and South America, Southeast Asia (the Philippines); Spanish in 1914: Europe; France was confined to Europe in 1714 and expanded colonial holdings over next two centuries. Spain, on the other hand, held colonies in both the Americas and Asia in 1714, but by 1914 had retreated back to Europe.

Review Exam

1. A global geography framework refers to a conceptual approach to studying world regional geography that includes a sustained and focused analysis of global interdependence, ongoing global and local change, and systems political-economic relations.
2. The paleolithic (or old stone age) refers to the period of time from about 4 million B.C.E. through 12000–10000 B.C.E. During this time, the majority of the world's human populations engaged in hunting and gathering for survival. The Neolithic (or New Stone Age) refers to a period of time where dramatic change in the social and spatial organizations of human life was taking place. During this period of time, humans began to adapt their environments more vigorously, developing both sedentary agricultural practices and systems of herding called pastoralism.
3. Globalization creates interconnections between economic, environmental, political, and social processes. Currently, most people live in places that are connected to world markets for goods and services (e.g., finance) through the flows of international trade and cultural exchange.

4. “Development that meets the needs of the present without compromising the ability of future generations to meet their own needs.”
5. GDP is an estimate of the total value of all materials, foodstuffs, goods, and services produced by a country in a particular year. GNP includes the net value of income from abroad, including profits and losses from overseas investments.
6. The demographic transition is, in part, the result in the changing economic base of a country/region/place from preindustrial to industrial. As this movement in time occurs, populations initially boom. As industrial culture emerges, birthrates decrease and the population appears to stabilize.
7. Some countries become stalled in a period of high-growth rates in population. These high growth rates are attributable to increases in medical technologies that have reduced infant mortality while cultural attitudes toward large families lag behind such innovations. The result is larger families and increased populations.
8. The experience of everyday routines in familiar settings allows people to derive a pool of shared meanings.
9. Regional geography combines elements of human and physical geography; regional geography focuses on understanding human–environment relations through a regional context; regional geography examines how places combine to construct regional identities. Human and physical geography have historically been systematic fields that focus on one aspect of human or physical experiences, such as cultural geography, fluvial geography, economic geography, etc., across space.
10. Spatial justice is a term used to represent the fairness of geographical variations in people’s levels of affluence and well-being, given people’s needs and their contributions to the production of wealth and social well-being.
11. Land masses have moved relative to each other and across Earth’s surface over millions of years. He determined this by showing how fossils from Africa and South America were similar and the fit of South America and Africa as continents along their coastlines.
12. Biodiversity: differences in the types and numbers of species in different regions of the world. Biodiversity is usually related to temperature. Since a large portion of the world exists in the tropics, we find high levels of biodiversity in those regions.
13. Answers will vary.
14. Many countries globally trade with over 120 different countries, demonstrating high levels of global interconnection. Despite that trend, there are regional disparities. Sub-Saharan Africa, on the whole, has far fewer trading partners than does the former French colonies of Southeast Asia. Mongolia has relatively little external trading activity as well.
15. Population projections for 2025 and 2050 worldwide show dramatic disparities in the growth of world populations. It is clear, for example, that while Europe, North America, and Latin America and the Caribbean will demonstrate mild to little growth, Asia in general and China and India in particular will see dramatic increases. It is predicted as well that Africa will continued to grow, becoming one of the most populous spaces in the world by 2050.
16. Nation is a term that denotes a people that recognize a common identity, but they may not reside in one common geographical area. Nation–state is an ideal form of a statehood under which everyone has a common nationality. Nationalism is the feeling of belonging to a nation as well as the belief that a nation has a right to self-determination. Nationalism plays the role of fostering the desire to construct nation–states.
17. The geographies of the Cold War were premised on a bipolar world, whereby many states aligned themselves with the dominant “superpowers” of the time, either the United States or the Soviet Union. Taking a view from the north pole, one sees the proximate relations that distinguished the extent of these two alliances. A series of proxy wars between these two alliances led to violence in Southeast Asia, Latin America, and Africa.

18. International division of labor refers to the process by which different people, regions, and places specialize in particular economic activities. Under colonization, colonial powers organized particular economic activities in each colonial holding based on local comparative advantages.
19. Answers will vary.
20. Colonization facilitated the movement of new foodstuffs globally. The potato, which was first domesticated in the Americas, was transplanted to Europe where it became a staple crop in a number of countries, including Ireland.
21. This change will be driven by industrialization targeting growing worldwide consumer demands. Both the manufacturing of those goods and the consumption of them after production will be increased dramatically in the global south, which have much larger populations.
22. Tropical climates can be found between the tropic of Cancer and the tropic of Capricorn on all continents. Dry and semiarid desert and steppe climates are also located throughout the tropics and also can be found in the midlatitudes. Mesothermal climates are found in parts of the southern United States, western Europe, east, southeast, and south Asia, South America, and the southernmost parts of Africa and Australia and New Zealand. Microthermal climates are located in the midlatitudes and high latitude regions of the Northern Hemisphere. Polar climates are found near both poles, while highland climates are most discernable at the global scale in the Himilayas, Rocky Mountains, Andes Mountains, and Ethiopian highlands.
23. In the tropical regions one would find tropical rainforests and savanna. In dry arid and semiarid regions there are mostly deserts (less than 10 inches of rain per year) and steppes. Mesothermal climates produce a variety of subclimates that sustain various forest lands and different types of vegetation. Polar and highland climates support little to no vegetation and, at best, are home to grasslands, and shrubs.

24. b	25. d	26. c	27. b
28. c	29. d	30. d	31. d
32. b	33. a	34. b	35. c
36. d	37. d		

Chapter 2: Europe

Vocabulary Review

1. northwestern uplands; north European lowlands; central plateaus; Alpine system
2. fjords
3. polder
4. acid rain
5. buffer zone; satellite
6. state socialist
7. balkanization
8. enclaves
9. ethnic cleansing
10. xenophobia
11. disinvestment; deindustrialization
12. Golden Triangle
13. fascism
14. welfare state

15. Pastoralism
16. tundra
17. steppe
18. latifundia; minifundia
19. entrepôt

Activities & Problems

1. (a) Northwestern Uplands
   (b) North European Lowlands
   (c) Central Plateau
   (d) Alpine System
2. Turkish; Algerian
3. Portugal; Spain; Italy; Yugoslavia; Greece; Ireland; Finland
4. Answers may vary. Some sample responses include: neocolonial connections between former colonies and European colonial powers; uneven development within Europe, whereby southern Europe has developed at a slower rate of growth than northern Europe; increased fluidity of borders because of the development of the European Union.
5. WW I and WW II both impacted birth and mortality rates, meaning that there are deficits in both men and women who were in Germany during those wars. A baby boom ensued after WW II, and then fertility rates slowed in the 1970s to the present day.

Review Exam

1. Trade with non-European countries allowed the development and refinement of several key technologies that expanded European interest in global travel. The discovery of gold and silver prompted the Spanish to further expand their trading networks and empire into both the Atlantic and Pacific regions.
2. A national economy in which all aspects of production and distribution are centrally controlled by government agencies
3. EU policy on agriculture had the effect of stabilizing agricultural prices and subsidizing farmers' incomes. These policies have led to a general trend away from mixed farming toward monocropping. Farm modernization has had some detrimental impacts, including increased exploitation of moorlands, woodlands, wetlands, and hedgerows. Additionally, there have been farm surpluses resulting from subsidization of farming.
4. Geographic situation provides access to southern and central Europe; home to former colonial power capital cities; it includes the industrial heartland of western Europe; and concentrated market of skilled labor and affluent consumers.
5. An emerging secondary core in Europe that stretches through a number of distinct physical geographic and cultural landscapes. There is a high rate of urban development, benefiting from deindustrialization in the Golden Triangle region that has pushed industry into the Southern Crescent, creating flexible production regions that allow for local areas to develop their own industrial culture.
6. Most Muslims in Europe can be found in the industrial cities of western Europe. France's Muslim population is predominantly from Algeria, Morocco, and Tunisia. Germany's Muslim population is mostly Turkish, while the United Kingdom's Muslim population is from East Africa and South Asia. These patterns partially reflect former colonial relations, particularly in the case of France and UK, while all modern-day migrants are driven largely by the post-WW II economic boom. Many Muslims came to fill jobs in the industrial sectors of western Europe, jobs that could not be filled by Europeans themselves.

7. a	8. d	9. b	10. d
11. c	12. b	13. b	14. a
15. d	16. a	17. F	18. T
19. T	20. F	21. F	22. F
23. T	24. T	25. T	26. T

Chapter 3: The Russian Federation, Central Asia, and the Transcaucaus

Vocabulary Review

1. tundra
2. taiga, tundra
3. chernozem; Siberian
4. unitary; federalist
5. pluralist democracy
6. civil society
7. permafrost
8. state socialism
9. dry farming
10. salinization
11. territorial production complexes
12. Silk Road
13. irredentist; secessionist
14. Eastern Europe
15. Ural Mountains
16. Black Sea
17. oligarchs
18. apparatchiks
19. Commonwealth of Independent States
20. Near Abroad

Activities & Problems

1. (a) 16th
   (b) 17th
   (c) 17th
   (d) 18th
2. (a) Estonia, Latvia, Lithuania
   (b) Poland, Hungary, Romania
   (c) 1945–1955

(d) 1946–1961

(e) Poland, Czechoslovakia, Germany

3. (a) Amur

   (b) Murmansk

   (c) Barents Sea; Kara Sea

   (d) Answer may include: Barents Sea; Kara Sea; Sea of Okhotsk; Sea of Japan; Pacific Ocean

4. Tundra: sparse vegetation with lots of rocky soil; Taiga: boreal coniferous forest; Steppe: tall, luxurious feather-grasslands with rich chernozem soil from high levels of decay.

5. The geographies of secession and irredentism in the region are complex. Secessionist movements and irredentist claims over territory in the region are driven, in part, by the ethnic geographies underlying the political geography of this region. A majority of these claims and tensions exist along the southern regions of the former Soviet Union, where new states emerged based on old Soviet provinces after 1989. In some of these states, large ethnic Russian populations remain, complicating the regional political geography of this area further.

Review Exam

1. Extensive plains interrupted by the Ural Mountains. A mountain rim creates a physiographic border that stretches from the Iranian border to the Sea of Okhotsk. The plains regions are bordered to the south by the Central Asian Plateau that extends from Iran to the Balkans. The northernmost part of the region is marked by permafrost.

2. Stalin withdrew from the capitalist world economy and relied on vast Soviet resources to develop a highly centralized, bureaucratic, industrial state. Stalin pushed for the collectivization of agricultural production, and capital for creating industry was to be extracted from agriculture. Stalin relocated peasants onto collective farms. In order to put the plans into action, Stalin relied on high levels of repression and exploitation. Despite the high levels of repression, the Soviet economy boomed under Stalin, growing at 10% per annum. There were severe environmental impacts from rapid industrial development as well, since the system was largely unregulated.

3. It is argued that the future of this region might take many different paths because on the one hand, global market forces could put pressure for more open and progressive economic and political systems in the region. On the other hand, further instability and reliance on primary commodities, could increase the social and spatial inequality in the region. Furthermore, the region's social and physical infrastructure will need to be enhanced to meet further development needs and interests.

4. The ongoing conflict in Chechnya is tied to this region's long-standing interest to remain independent from the Russian Empire, the Soviet Union, and now the Russian state. They declared independence in 1989 but the Russians invaded in 1994 to secure the country for Russia. Despite quasi-independence for Chechnya, in 1999, Chechen rebels attacked Russian cities. Russia invaded and occupied the country's capital until April 2009. Thousands of people have been displaced and torture and violence have been documented by human rights organizations against both the Russians and Chechens.

5. Some of the causes of crime and corruption during the period of market reform include: no accepted codes of business behavior, no civil code, no effective banking system, no effective accounting system, disorganized security agencies, porous borders, and poorly organized bureaucracies. This fostered the emergence of ethnically-based *mafiyas*.

6. a
7. b
8. d
9. c
10. b
11. a
12. a
13. b
14. a
15. a
16. F
17. T

18. T
19. F
20. F
21. T
22. T
23. F
24. T
25. F

Chapter 4: Middle East and North Africa

Vocabulary Review

1. aridity
2. oasis
3. without irrigation
4. afforestation
5. Ottomans
6. Balfour Declaration; Palestine
7. informal economic activities
8. import substitution; nationalization
9. submission; Muslim
10. Islamism; anticolonial/anti-imperialist movements
11. jihad
12. kinship
13. subsistence activity that involves breeding and herding animals to satisfy human needs for food, shelter, and clothing; transhumance
14. chador; veil
15. guest workers
16. hajj
17. intifada
18. Petrodollars
19. Zionism
20. internally displaced person
21. desertification

Activities & Problems

1. (a) French
   (b) British
   (c) Italian
   (d) British/Ottoman
   (e) French
   (f) British/Ottoman
2. (a) Saudi Arabia, Iran
   (b) Tunisia, Turkey, Israel, Bahrain
   (c) Algeria
   (d) Turkey, Israel

3. The migrations are into the oil-producing states, such as Saudi Arabia and Libya. There are also migrations for political reasons, such as out of western Sahara. Migrations began in the 1950s and were largely prompted by the development of oil economies in the Gulf States.
4. Israel in 1949 consisted of the older territory of Palestine except for the West Bank and the Gaza Strip. In 1967, Israel occupied Gaza, West Bank, Sinai Peninsula, and Golan Heights. In 1996, Israel remained in all the territories taken in 1967 except Sinai. Today, Gaza is under Palestinian control, as are parts of the West Bank.
5. Answers will vary.

Review Exam

1. A majority of the region is marked by dry arid/semiarid climates, although there are also midlatitude climates found along the north coast of Africa and along the Mediterranean coast of Israel, Lebanon, Syria, Turkey. Midlatitude climates structure into the Zagros Mountains of Iran. In southern Sudan there are both tropical climates along the border with the Central African Republic and Uganda and highland climates along the border of Ethiopia.
2. Plant and animal domestication facilitated the movement to agriculture, while the production of staple crops eventually led to the development of village-based economies. Populations slowly increased as food surpluses were produced, and social organization changed as people moved from relatively egalitarian nomadic groups to villages based on kinship networks. Leisure time increased and specialization in nonagricultural crafts was possible, giving way to new technological and social innovations. The beginning of more complex barters and trade systems developed.
3. Transhumance means the movement of herds according to seasonal rhythms. Transhumance differs from sedentary life because people move more regularly and are dependent on the collection of wild grains and on the occasional slaughter of their herds for food, clothing, and shelter. In some cases, women would be in charge of planting grains during the spring seasons.
4. Colonial politics in the region, particularly in the 1910s, established the template for a Jewish homeland in the region under British control. British control over the region fosters Jewish ingration, while UN support lends credence to an independent Jewish state in the region. Population changes followed throughout this period, with large scale Jewish inmigration, which increased after independence, with the Law of Return. Palestinians, on the other hand, were displaced. The wars of the 1960s-early 21st century have further changed the population geography of the region, as Palestinians were further displaced as political refugees, isolated either in camps outside of Israel (e.g., Lebanon), in new autonomous territories (e.g., Gaza), or further abroad. Despite all this, the population of Israel remains somewhat evenly divided between Jews and Arabs by around 2005.
5. It is clear that the future of water and oil politics in this region will become further complicated over time. First, international pressure on the burning of fossil fuels threatens to dampen the economic benefits of oil production in the region. Second, water remains a scarce commodity in the region, with many countries turning toward desalinization to meet their growing water needs. Third, despite these problems, this region will continue to garner global attention as regional conflct threatens to cut off needed supplies of global oil production.

6. d	7. c	8. d	9. b
10. c	11. a	12. a	13. d
14. a	15. c	16. T	17. T
18. T	19. F	20. T	21. F
22. T	23. F	24. T	25. F

Chapter 5: Sub-Saharan Africa

Vocabulary Review

1. harmattan
2. domestication
3. blood diamonds
4. scramble for Africa; Berlin Conference
5. apartheid
6. homelands
7. IPCC; Intergovernmental Panel on Climate Change
8. feminization of poverty
9. gender and development
10. microfinance; programs
11. Horn of Africa
12. slash and burn
13. Afrikaans
14. environmental determinists
15. hut; indirect
16. Royal Geographical Society
17. Millennium Development Goals
18. lingua franca
19. modernization; dependency
20. urban bias

Activities & Problems

1. (a) Central Africa, Southern Africa
   (b) West Africa, Central Africa, Southern Africa, North Africa
   (c) North Africa
2. (a) Nigeria
   (b) Algeria
   (c) some combination of Mauritania, Sudan, Equatorial Guinea, Angola, Democratic Republic of Congo, Gabon
3. (a) 1895; 1908; 1899; 1908
   (b) German East Africa; German Southwest Africa; Cameroon; Togo
   (c) Angola; Mozambique; British
   (d) Portugese; Belgian; British; French; German
4. From the two graphs, it is clear that increases in education lead to decreases in fertility in all three countries. As the caption suggests, South Africa has the highest per capita GNP of all three countries, suggesting that there is also a relationship between GNP per capita and fertility. The bottom graph also shows that fertility rates are higher in rural areas in all three countries.

5. Throughout the central region there is tropical rainforest indicating high levels of rainfall. The tropical rainforest does not make it to the east coast, probably due to the mountains of the Great Rift Valley. Just north and south of the tropical climate are bands of savanna, indicating less rain. Continuing north and south there are deserts and then finally Mediterranean climate regions. Along the Great Rift Valley there are also highland climatic regions. In terms of vulnerability to desertification, many parts of the continent just south of the Sahara (in the area known as the Sahel) are vulnerable to deserfication, as are areas just north of the Sahara along the coast. Additionally, parts of east Africa are also at high risk for desertification.
6. Cell phone coverage is most dense in South Africa and in parts of east Africa, with strong concentrations of cell usage along the western and northern coasts. According to the figure, cell phone use is highest in Gabon (87%), South Africa (87%), and Botswana (60%).
7. Answers will vary.

Review Exam

1. There are five major basins located across the continent in the western, central, and southern regions. Mountain ranges can be found across the northern region known as the Maghreb, the Ethiopian Highlands in the Horn of Africa, and bordering the Great Rift Valley. Mountain ranges also frame the southern part of the continent on both the east and west coasts. Several major rivers and lakes have left marks on the landscape, creating the Nile and Congo valleys, for example.
2. According to all four maps, every country (except Somalia and Djibouti, where no data are available) has already reported AIDS cases. Over time, the southern and central regions of the continent have seen higher percentages of adults with AIDS.
3. This region is called the "cradle of mankind" because there have been the earliest signs of the development of human beings in this region. Indications of early ancestors of humans were found to be from 3.7 million years ago.
4. During the height of the Cold War, many countries in Africa were experiencing independence movements and transitions to postcolonial states. Leaders in various countries were exposed to socialism as well as communism. Both the Soviet and American blocs supported various factions as tensions mounted over control of the newly formed states. Examples include the civil violence in Angola.
5. The colonial period left many countries in the region underdeveloped politically. Politically, weak local social infrastructures have meant that numerous factions have been vying for control, and many countries are still marked by civil unrest and political instability.
6. Economies in this region have grown rapidly since the 1990s, averaging growth around 5% per annum. Cell phone usage is also up, as is access to cell phone networks, suggesting that change has occurred. Challenges remain, however, including the ongoing problems related to armed conflict, stabilizing illness and death related to epidemic diseases, stabilizing food security.

7. c	8. a	9. a	10. c
11. a	12. d	13. c	14. a
15. d	16. T	17. T	18. F
19. F	20. T	21. F	22. T
23. T	24. T	25. T	

Chapter 6: The United States and Canada

Vocabulary Review

1. intermontane
2. europeanization

3. americanization
4. indentured servants
5. staple economy
6. hate crimes
7. airsheds
8. multiculturalism
9. internal migration
10. suburbanization
11. antiglobalzation
12. creative destruction
13. cap-and-trade
14. carbon footprint
15. Superfund sites
16. megalopolis; Main Street
17. GMO; genetically modified organisms
18. organic farming
19. federal states
20. slow food

Activities & Problems

1. Marine coastal, Highland, Arid and semi-arid, Highland, Continental/mid-latitude
2. (a) Brazil, China, United States, Zimbabwe
   (b) China, Brazil, India, United States
   (c) Africa and the Middle East
   (d) Zimbabwe, Brazil, Pakistan
3. (a) The spatial pattern of total emissions across the U.S. shows that the density of emissions is concentrated in the eastern U.S. as well as in the western U.S., with decreasing intensities throughout the central region of the country. There are clearly urban densities across the country as well, where emissions are intense and networks (likely towns connected via roads) also generate higher emissions across the country.
   (b) Population density appears to be much higher in the eastern half of the United States, with population rates dropping preciptously through the central region of the country. Population centers increase in the western United States. Densities do vary, however, in the eastern U.S., which is likely a result of urbanization.

   There is a clear spatial relationship between the pattern represented in Figure 6.4 and 6.18, with emissions focused around urban nodes in the eastern U.S. .
4. (a) Large tracts in the South and the Southwest demonstrate high poverty only, while the region of megalopolis and California have large areas of high wealth only. Most of the country is home to areas of high poverty and wealth. Very few places are sites of high equality, although some of the places can be found scattered throughout the Midwest.
   (b) Answers may vary.
5. (a) The pattern demonstrates the three waves of migration to the U.S.: (1) 1820–1870, large migration stream from Northern and Western Europe; (2) 1870–1920, increase in Southern and Eastern

European migrants accompanying Western and Northern Europeans; (3) 1970–present, Asian and Latin American migrants.

(b) The most recent wave of migration is dominated by peoples from Latin America, particularly people from Mexico. Reforms on strict immigration legislation in 1980 fostered an increase in Mexican migrants to the U.S. where they now make up 7.3% of all foreign-born peoples in the country. Many come to the U.S. seeking work as migrant laborers or to work in factories.

(c) The reduction in in-migration between 1930 and 1940 marks the high point in the U.S.'s strong anti-immigration stance, which was precipitated by a backlash against southern and eastern Europeans by the majority western and northern European populations that inhabited the U.S. at the time. Immigrants were seen as dangerous to U.S. values. The U.S. imposed race-based quotas to also reduce the reduction of Asians, particularly Chinese, migrants to the country.

Review Exam

1. Main approach: identify areas of high potential for renewable resources with low environmental impact and greenhouse gas emission potential. These energy sources include wind, solar, hydropower, and geothermal energy.

2. Distributions of Native American subsistence practices were significantly related to the local climatic and vegetative geographies of the region. In the northernmost parts of the continent, Native Americans engaged in Arctic and sub-Arctic hunting and fishing. Moving south into the northeast part of the modern-day United States, Native Americans engaged primarily in woodland farming. In the southern part of the United States, they were also farmers. West into the Great Plains, hunting became the primary subsistence activity. In the Rocky Mountain region, Native Americans engaged in hunting and gathering, while along the coast fishing was the dominant mode of subsistence. Finally, in the Southwest and modern-day Mexico, Pueblos practiced sophisticated farming techniques that had been passed down for millennia.

3. Two of the most long-term and successful social movements in this region has been the women's and environmental movements. The women's movement pushed hard for women's voting rights in both countries in the early 20th century, continuing to expand on the rights of women (including over their own body) throughout the century. The environmental movement became significant in the 1800s, when major figures began to lecture on the need to protect the environment. National parks were established and continue to play an important role in protection of the environment in the region. Recent moves toward green technology has also helped facilitate a wider conversation about conservation in the region. Antiglobalization movements have also been firmly established in both countries, with groups and individuals challenging the dominance of global corporations and international trade agreements in the regulation of everyday consumer practice.

4. The Québec movement for independence has failed twice but the desire to maintain the individual identity of the people in that region remain. Recently, the province has focused its energies on increasing its connections to Europe by taking advantage of the North American Free Trade Agreement.

5. The cities of San Francisco, Los Angeles, and San Diego, as well as Sacramento play an integral role in the expanding power of the Pacific Rim region, which includes key parts of Asia as well as Vancouver in Canada. The development of the Asia-Pacific Economic Cooperation in 1989 is now an organizational force in discussions about how to improve economic relations among member nations. As a result of these connections, the new economy, based in the information and technology industries, brings together the economies of Silicon Valley in California, Tokyo, Japan, Singapore, and Vancouver, Canada, to name a few.

6. d	7. d	8. c	9. a
10. b	11. d	12. b	13. b
14. b	15. d	16. T	17. F

18. F
19. F
20. T
21. F
22. T
23. F
24. T
25. F

Chapter 7: Latin America and the Caribbean

Vocabulary Review

1. Altitudinal zonation
2. pristine myth
3. Treaty of Tordesillas
4. demographic collapse
5. haciendas
6. plantations
7. altiplano
8. structural adjustment policies
9. Mulatto; zambo
10. urban primacy
11. Braceros
12. liberation theology
13. favelas
14. nontraditional agricultural exports
15. informal economy
16. Columbian Exchange
17. ecotourism
18. bioprospecting
19. Monroe Doctrine
20. offshore financial services
21. remittances
22. free trade zones
23. contract farming
24. circle of poison

Activities & Problems

1. Tropical desert, highland climates, midlatitude seasonal climates, tropical savanna climates, midlatitude seasonal climates, tropical savanna climates, tropical rainforest
2. (a) Answers include any three of the following: sugarcane, bananas, yams, maize, rice, poultry, pigs, and cattle.
   (b) Answers include any three of the following: coffee, maize, coca, tomatoes, melons, cut flowers, dairy products, and cattle.
   (c) Answers include any three of the following: wheat, barley, maize, potatoes, apples, sheep, guinea pigs, llamas, alpaca, and vicuña.
   (d) Answers include any three of the following: highland grains and potatoes, sheep, guinea pigs, llamas, alpacas, and vicuña.

3. (a) 1981–1988
   (b) 1973
   (c) 1982
   (d) 1983
   (e) 1954
4. (a) Portugese
   (b) Spanish
   (c) French
   (d) English
   (e) Dutch
   (f) Mayan
5. Mayan: Yucatan Peninsula and south into present-day Guatemala; Aztec: central and parts of southern Mexico stretching from the Pacific to the Caribbean; Inca: west coast of South America from the southern border of present-day Colombia to central Chile (dominated the Andes).
6. (a) South America's interior is dominated by the Amazon Basin, while the west coast of the continent consists of the Andes Mountains and a small coastal plain. North and south of the Amazon Basin are two highland ranges. South of the Brazilian Highlands is the area known as the Pampas, a relatively flat alluvial plain. Much of Central America is mountainous, and mountain chains surround central Mexico, leaving only a small coastal plain on the western side of the country. The Yucatan Peninsula is relatively flat. The Caribbean is a series of island chains that surround the Caribbean Sea.
   (b) Mountain ranges, such as the Andes, have resulted in dramatic differences in climate. These mountains, while situated in the tropics, still have snow pack. The Amazon Basin, on the other hand, is a flat region containing almost 20% of the world's freshwater. It is home to a combination of grasslands and rainforest. Volcanic activity is a staple of plate boundary countries, such as those in the Caribbean. Sharp peaks and tectonic activity mark many parts of the region. Because of the complex tectonic geographies of the region, the region is rich in numerous mineral resources.
   (c) Mineral wealth can be find throughout the old crystalline shield where the crust has folded, bringing older rocks to the surface of the earth. The Andes regions of Peru and Bolivia remain important mining regions. The region is also known for its rich oil, gas, and coal deposits, based, in part, on the much longer geologic history of the region.
   (d) El Niño brings warmer, wetter winds to the coast of South America, a normally dry desert-like region. The increase in rains can cause flooding and high levels of rainfall. The result is sometimes global drought conditions. La Niña, on the other hand, creates rains in northeast Brazilian, drought in Mexico, and hurricane conditions in the Pacific.
   (e) Biodiversity: high levels of biological uniqueness and variation. This particular region has high levels of biodiversity because of the tropical climates, which are able to sustain significant variations in biological organisms. Consistent temperatures and moisture facilitate the growth of myriad flora and fauna.

Review Exam

1. Theories for the decline of the Maya include overuse of lands that led to environmental degradation and social collapse as the state could no longer support large populations.
2. The Incas adapted to their environment by building terraces throughout the mountainous region. Terraces collected water and reduced soil erosion, while also providing a flat surface on which to plant. Moreover, the terraces reduced frosts by reducing downhill flows of cold air.

3. The stark differences between rich and poor have their roots in the colonial infrastructure and the hacienda system. Under that system, a minority was given control over land and resources. In the postcolonial period, oligarchies rose based on the hacienda system, and wealth continues to be concentrated in the hands of a few. The results have been rapid urbanization and poverty in both rural and urban areas as corrupt politics maintain high degrees of economic difference between rich and poor. In some cases, a middle class has emerged, but in many other cases, a two-tiered system remains.
4. Liberation theology comes out of the Catholic tradition and is focused on developing the economic, political, and social capabilities of the poor. It is a combination of the philosophies of Jesus Christ and Karl Marx, both of whom spoke about inequality and oppression. Priests developed grassroots self-help groups, or what are often called Christian-based communities. They often spoke out against repression and authoritarianism.
5. The geologic history of the region created considerable mineral wealth from precious and industrial metals (gold, silver, copper, tin, aluminum, etc.) to oil and natural gas. Mexico has been drilling oil since the 1890s and Venezuela was one of the founding members of the Organization of Petroleum Exporting Countries (OPEC) in 1961, but fossil fuel deposits extend throughout northern South America. Colombia and Brazil have not effectively exploited their petroleum reserves for political and environmental reasons.
6. The economic future of this region illustrates how important broader global relations are in terms of economic development. Markets beyond the borders of this region, particularly in the United States, may remain unstable, therefore impacting the abilities of this region to export its primary and secondary commodities effectively. The interdependence of this region, then, with other world regions cannot be underestimated. In terms of human-environment relations in the region, the push toward increasing industrial development will mean that governments in the region need to consider the impacts this might have on the pollution emissions in the region. Growing populations and new consumer habits will also impact how the valuable resources and biodiversity of this region is used in the future. Effective management in the wake of uncertain economic times will be challenging.

7. b	8. c	9. c	10. c
11. a	12. b	13. c	14. a
15. d	16. c	17. T	18. T
19. F	20. T	21. T	22. T
23. T	24. F	25. F	

Chapter 8: East Asia

Vocabulary Review

1. zaibatsu
2. Pacific Rim
3. Ring of Fire
4. Anime
5. Keiretsu
6. Asian Tigers
7. counter-urbanization
8. feng shui
9. hydraulic civilizations
10. Great Leap Forward
11. chaebol

12. massif
13. pinyin
14. ideograph
15. Shinto
16. Special Economic Zones; SEZs
17. import substitution
18. one-child policy
19. Confucianism
20. manga

Activities & Problems

1. (a) arid and semi-arid
   (b) continental/mid-latitude
   (c) humid subtropical
   (d) continental/mid-latitude
   (e) monsoon
   (f) highland or humid subtropical
2. (a) Main islands of Japan, Formosa (Taiwan), Korea (North and South), Sakhalin Island
   (b) Expanded empire to include Manchuria, a region whose northernmost extent is the Soviet border.
   (c) In 1933, Japan extended its sphere of influence into the Great China Plain and almost to Beijing. They also began to assault China through the port city of Tsingtao.
3. Population density demonstrates a significant decrease moving east to west. Japan, Korea, Taiwan, and the coastal plain of China have very high population densities, while the Tibetan Plateau, Mongolia, and the region of the Taklamakan Desert have much lower population densities. Resource allocation in the region would have to be toward areas of high density, including the urbanized parts of the region.
4. (a) United States
   (b) Australia, Cuba, Peru, Mexico
   (c) United Kingdom, France
5. (a) East Asia is divided into three distinct physiographic regions: the Tibetan Plateau, the central mountains and plateaus, and the continental margin. The climatic geography of this region is partially influenced by that tripartite division, with highland climates found in the Tibetan Plateau region with arid and semi-arid regions found north of that plateau in Tibet's rainshadow. In the contintental margin, we find two major climate regions-humid subtropical in the south and continential/mid-latitude in the north. The vegetation of this region is affected by these two geographies—with grasses and drough resistant plants found in the high plateaus, vegetation that is both drought and salt resistant in the arid/semi-arid spaces of the region, and more water-dependent vegetation in the continental margin, including temperate broad-and needle-leaf trees.
   (b) Tibetan Plateau: grazing; Mongolia: pastoralism; North China Plain: wheat and barley cultivation; southern China: rice and tropical fruits.
   (c) Much of the land in the flat plains must be used for producing staple crops, such as rice and wheat, and this puts more pressure on land and population density.

(d) Answers will vary.

(e) Answers will vary.

Review Exam

1. Unified China under an imperial bureaucracy created stability that eventually facilitated the development of the Silk Road. To protect the empire, the Qin started building the Great Wall.
2. The Asian Brown Cloud is a blanket pollution nearly 2 miles thick that sits over parts of the Indian Ocean, Southeast Asia, and East Asia. It consists of sulfates, nitrites, black carbon and other pullutants, a result of the increased use of fossil fuels in the region.
3. Tensions exist in Tibet and in westernmost China. Tibet was invaded by the communist Chinese regime in 1950 and has been under "occupation" ever since. Tibet is predominantly Buddhist, and the Han Chinese wanted to control religion in the region. Today, temples are closely monitored by Han Chinese/communist officials. In westernmost China, religion is also a major factor. In this case, however, it is a suppressed Muslim population that has pushed for greater autonomy. Many of the people of this region are also nomads, who came under the jurisdiction of strict mobility policies under the communist regime. Violence broke out in this region in July 2009.
4. The Great Leap Forward was a scheme to rapidly increase economic productivity in China in the late 1950s. Land was merged into communes with the goal that industrialization would occur throughout the country in what came to be known as "backyard furnaces." It had a dramatic impact on the landscape and on everyday life. Large-scale communes replaced historically smaller farming plots, and the historical pastiche of multiple crops was replaced by monocropping. There was no planning on consumption, however, and industrialization in the countryside failed. Bad weather coupled with planning failures led to famine conditions, and 20 to 30 million people died.
5. Under the liberalized policies of Deng, migration from rural to urban regions increased dramatically, putting pressure on many urban areas. In the 1990s, the percentage of the population living in urban areas doubled. To alleviate stress on China's largest cities, the government created a series of buffer cities to absorb the flow of migrants. One of the major consequences of the increase in rural-to-urban migration has been a loss of arable land to housing, roads, and factories.

6. c	7. d	8. a	9. d
10. b	11. a	12. b	13. c
14. d	15. c	16. F	17. F
18. T	19. T	20. F	21. T
22. T	23. T	24. T	25. F

Chapter 9: South Asia

Vocabulary Review

1. Raj
2. distributaries
3. Mughal
4. Hinduism
5. Bollywood
6. Caste
7. Green Revolution
8. Hindutva
9. orographic effect

10. Sundarbans
11. bonded labor
12. terracing
13. zero population growth; ZPG
14. Grameen Bank; microfinance

Activities & Problems

1. (a) Ladakh, Kashmir and Jammu, Belutshistan, Rajputana, Fatehpur, Nitzam's Dominions, Mysore, Bhutan, Carchar, Manipur

   (b) 1753–1805: annexation took place along the eastern coast and up through the Ganges and Brahmaputra river valleys. After 1805: further extended control into the central regions of the country, north into present-day Pakistan and east into the northeast frontier and Assam.

   (c) Afghanistan and Nepal

2. Hinduism: Most of India, north coast of Sri Lanka. Islam: Pakistan, Bangladesh, and pockets throughout India, most highly concentrated along the borders with Pakistan and Bangladesh. Buddhism: Sri Lanka, parts of Nepal, Bhutan, and parts of northern India near Jammu and Kashmir. Christianity: east of Bangladesh in India, pockets along west coast of India and a few pockets along east coast of India, as well. Sikhism: north of New Delhi in India along the Pakistan border.

3. (a) Afghanistan

   (b) Afghanistan

   (c) Afghanistan, Nepal, India

4. (a) India is surrounded by a coastal fringe that gives way quickly to a series of mountains. The interior of India is distinguished by the Deccan Plateau, which extends to the Ganges and Bhramaputra River valleys. Surrounding the Ganges River valley are plains that eventually meet up with the "mountain rim," which includes the Hindu Kush and Himalayan Mountains, and the Khasi and Chttagong hills.

   (b) 1. Plains

      2. Peninsular Highland

      3. Mountain Rim

      4. Coastal Fringe

      5. Mountain Rim

5. (a) The Soviet's believed that the instability in Afghanistan posed a serious security problem along its southern border and that insurgency had led the government to be less interested in acting as a satellite of the Soviet regime.

   (b) Civil war ensued after the Soviets pulled out in 1989. Civil war erupted between the *mujahideen* militias vying for control of the war-torn country.

   Moreover, the Taliban regime was attacked in the post-9/11 period, as the U.S. saw this state as supportive of Al-Qaeda and global terrorism. Although the Taliban were pushed to the margins and Al-Qaeda all but driven out of the country, a resurgent Taliban has made strides recently in the country. (c) The presence of the U.S. in Afghanistan has caused increased tensions with Pakistan, which is likely the new home to the leaders of Al-Qaeda. As state control has been difficult to acheive, it is clear that U.S. and NATO forces will have to conduct military operations in the Afghanistan-Pakistan border region.

   (c) The presence of the U.S. in Afghanistan has caused increased tensions with Pakistan, which is likely the new home to the leaders of Al-Qaeda. As state control has been difficult to acheive, it is clear that U.S. and NATO forces will have to conduct military operations in the Afghanistan-Pakistan border region.

Review Exam

1. The plains are dominated by three main river systems: Indus, Ganges, and Bhramaputra, flowing over a region of young sedimentary rock. The Ganges and Bhramaputra meet and form a giant delta region. These rivers provide the region with a steady, yet uneven, amount of snowmelt. Irrigation in the plains allows for large-scale agriculture and high population densities. The coastal fringe, on the other hand, consists of a mixed number of physical features. This is a relatively narrow stretch of land marked by marine erosion that has exposed the edge of the ancient shield of the Peninsular Highlands. Some parts of the coastal fringe have alluvial deltas from water that runs down the Western and Eastern Ghat Mountains. The land is relatively fertile and has supported population densities, which are not as high as the plains region.
2. Summer monsoons bring rain from the Indian Ocean into the South Asian region, dropping particularly heavy amounts of rain in Bangladesh and Pakistan and along the western coastal plain. In the winter, the winds switch and bring rain to the east coast of Sri Lanka and parts of the southern eastern coastal fringe. The impact has been to sustain large-scale wet rice agriculture (along with other export crops) in the rich delta plains of the Ganges and Bhramaputra rivers.
3. Colonial legacy, language, and religion are the factors that fuel tension in Sri Lanka. Tamils, Hindus that speak Tamil, were brought to the island under British rule to work the fields and the bureaucracy. They were given disproportionate access to resources. The Sinhalese majority, who are Buddhists, resented the Tamil incursion, and under independent elections in the postcolonial period took control of the government. Tamils have since fought for complete autonomy for the northern region of the country where most Tamils live.
4. Based on its industrial geography, India's development remains uneven. Areas of industrial growth have seen much higher levels of economic growth and increasing rates of standards of living for the growing wealthy and middle classes. As an example, one can look at the "silicon valley" of India around Bangalore and Hyderabad. Areas with initial advantages in industrial development have received increased benefits, while areas of relatively low capitalist productivity have suffered, particularly rural areas.
5. Aryan peoples, who were nomadic pastoralists, migrated into the region between 1500 and 500 B.C.E. developing agriculture in the region of the Ganges River through a massive deforestation and irrigation effort. The spread of agricultural civilization spread throughout the region, with a majority of the region being farmed by small-scale farmers. This population also increased the ties between this region and other worlds, increasing trade links between the Ganges Plain and other parts of the world. Moreover, the Aryans helped expand urban life in the region.
6. While it is too early to tell exactly what impact climate change will have on the region, it is clear that there could be increased droughts and flooding, as rainfall patterns change over time. Higher temperatures have already damaged crops and food production as well in the region. In the Himalayas, glaciers are decreasing in size, causing concerns about water security in the region. Furthermore, increasing temperatures could have a dramatic impact on the disease ecology of the region, with new places and peoples becoming susceptible to diseases that weren't near their homes before.

7. d	8. a	9. c	10. a
11. c	12. b	13. b	14. b
15. d	16. T	17. F	18. T
19. T	20. T	21. T	22. F
23. F	24. T	25. F	

Chapter 10: Southeast Asia

Vocabulary Review

1. monsoon
2. Wallace's line

3. sawah
4. spice islands
5. culture system
6. transmigration
7. Overseas Chinese
8. world city
9. Mekong River Commission; MRC
10. green revolution
11. tsunami
12. the Association of Southeast Asian Nations; ASEAN
13. domino theory
14. killing fields
15. Free Trade Zones
16. hill tribes; chao khao
17. Swidden
18. crony capitalism; kleptocracy
19. AIDS orphans
20. overseas chinese; rice bowl
21. tsunami
22. domino theory
23. matrilocality; patrilocality
24. Culture System
25. Bumiputra

Activities & Problems

1. (a) tropical savanna

   (b) tropical wet with no dry season

   (c) monsoon

   (d) humid subtropical

   (e) Borneo; New Guinea (other correct answeres: Irian Jaya, Indonesia or Papua New Guinea)

2. Vegetation varies from monsoon forests sporadically located across mainland Southeast Asia to rainforest, also across the mainland and dominating much of the insular areas of the region. Monsoon forests can also be found sporadically located across parts of the insular region. Much of the mainland, the Philippines, Java, and Sumatra are marked by "other land use." Other land use stands in for agricultural land production. The map thus represents not only land cover but also land use. The river cultures of the mainland are marked by high degrees of agricultural productivity. Java is nearly completely deforested as are most of the Philippines. One can find secondary forest growth in parts of the Philippines and Laos, indicating either afforestation projects or severe land degradation that led to abandonment and secondary growth.

3. (a) Indonesia

   (b) Brunei Darussalam

   (c) Thailand

(d) Myanmar

(e) Myanmar

(f) Vietnam

(g) Lao PDR; Malaysia; Thailand; Vietnam

4. (a) Laos, Vietnam, Cambodia

   (b) Burma, Malaysia, Singapore

   (c) Indonesia

   (d) Philippines

   (e) Philippines

   (f) 1570; 1898

5. Japan first invaded East Asia taking Korea, Manchuria. In 1941, Japan extended its holdings to central China, French Indochina, Mariana Islands, and Marshall Islands. In 1942, Japan extended its influence into all of Southeast Asia, including Indonesia, Burma, Thailand, and Papua.

Review Exam

1. The insular region is bombarded with rains from both monsoons, while the mainland is only significantly affected by one monsoon season.
2. The highest levels of deforestation have been witnessed in the mainland, the Philippines, and Java. All three areas have the highest population densities in the region. The regions that have witnessed the least are the lower population density areas, such as Borneo, the mountainous regions of Sumatra, Papua, and the northernmost parts of Burma.
3. The transmigration program was an action taken by the Indonesian government to move people from regions of high density to regions of lower density, specifically off Java and onto other major islands. Problems include lack of infrastructure for migrants, conflict between migrants and indigenous groups, destruction of forests for agricultural production, malaria, pests, and weed invasion with migrants.
4. Overseas Chinese played the role of entrepreneurs in the colonial economy. They established banks, insurance companies, and shipping and agricultural businesses.
5. The Middle East and East Asia have received the largest numbers of migrants from Southeast Asia. The region is also marked by relatively high levels of internal migration between ASEAN members. Internal migration is largely to Malaysia and Singapore. Women work as maids and domestics in the Middle East. Philippine men have joined the merchant marines. Thais and Filipinos also work in Hong Kong and Taiwan because they are paid less than local labor. Filipino women also work as maids in East Asia and North America.
6. Even though SEA is often discussed as a region with a few key ethnic groups, all countries in the region have minority populations. One of the major problems for ethnic minorities is the rights to citizenship. Many minority populations also suffer abuse at the hands of the majority population. This has created the consequence of refugees throughout the region. Migratory peoples, such as the sea gypsies are also marginalized, as governments try to regulate their movement through settlement programs.
7. HIV cases are highest in the country of Thailand, although rates are increasing in countries throughout the region, particularly in Burma, Cambodia, and Vietnam. In Thailand, an effective public health campaign has reduced rates of HIV, while in other countries poorer public health systems cannot effectively respond to the crisis. In many places, a lack of access to education and medicine means that there are high levels of discrimination. Moreover, under-reporting is common because of the stigma attached to HIV. There is a clear relationship between HIV rates and the extent of commercial sex work and drug use in the region. In areas throughout mainland SEA, one can find easy access to inexepensive commercial sex and drugs, particularly heroin but more recently methamphetamines.

8. c	9. a	10. b	11. d
12. c	13. b	14. a	15. b
16. d	17. a	18. F	19. T
20. T	21. F	22. T	23. T
24. F	25. F	26. T	27. T
28. T	29. F	30. F	

Chapter 11: Australia, New Zealand, and the South Pacific

Vocabulary Review

1. Antarctic Treaty
2. cattle stations
3. common property resources
4. ozone depletion
5. stolen generation
6. South Pacific Forum
7. Closer Economic Relations Agreement
8. White Australian policy
9. Dreamtime
10. Treaty of Waitangi
11. atoll
12. ecological imperialism
13. marsupials
14. island biogeography
15. guano
16. Micronesia
17. Melanesia
18. Great Artesian Basin
19. subsistence affluence
20. World Heritage Site

Activities & Problems

1. (a) monsoon, woodland
   (b) arid and semi-arid, shrub land
   (c) arid and semi-arid, shrub land
   (d) marine coastal, temperate forest
   (e) humid subtropical, rainforest
   (f) marine coastal, temperate forest
2. (a) Moving from east to west one finds a thin coastal plain, followed by mountain ranges, an interior lowland, then a raised plateau, with plains just to the south of the plateau and just west and north of the plateau.
   (b) High mountain ranges mark the center of both islands with narrow coastal plains.

(c) Answers will vary.

(d) Answers will vary.

3. (a) 1770

   (b) Hawaii

   (c) 1788; 1804; 1803; 1869; 1829

   (d) 1643; Fiji

4. (a) Uranium is an element found in nature that can be used effectively to create nuclear weapons. One can find uranium in northern Australia, which, as a country, contains almost 27% of the world's uranium.

   (b) It is significant because of the importance these islands have played in the advancement of nuclear technology globally. Uranium, and nuclear tests more generally, link this region to the global geopolitical landscape of the nuclear age.

   (c) Nuclear testing has created both political and environmental problems in the region. Environmentally, islands have been made uninhabitable and nuclear fallout has impacted surrounding populated islands. Politically, people have been removed from islands controlled by nuclear powers for testing, thus displacing their homes and cultural traditions.

   (d) Private organizations, such as Greenpeace, have tried to stop nuclear testing, while countries, such as New Zealand, have banned all nuclear-powered vessels from entering its ports. People have also rioted against the French in Tahiti.

   (e) The results have been the banning of nuclear testing in the region.

5. (a) –51 C (–60 F)

   (b) United States; Russia

   (c) 30,000

   (d) Chile

   (e) Petrels

Review Exam

1. Develop a more multicultural political and cultural system in both countries; increase indigenous landholdings; economic and social status of indigenous peoples. These issues have been treated differently in the two countries. New Zealand has developed a multicultural policy, whereby indigenous peoples are recognized and their rights/concerns addressed. At the same time, there is conflict over recent decisions over Maori representation on the Aukland Super City proposal. Australia, while granting certain concessions to indigenous peoples, still remains a place where indigenous peoples have less power and recognition. That said, recently the newly installed Labor Government in Australia did formally apologize for the atrocities against Aboriginal peoples.

2. Australia and New Zealand's economies were first based on wool production, although wheat farming also became important in southeastern Australia. The discovery of gold brought people in from across the globe, including Europe and China. Railroads and livestock yards facilitated the growth of the economy. And sugarcane was introduced into the northeastern tropical region of Australia. New Zealand's economic development mimicked Australia's.

3. The Treaty of Waitangi in New Zealand has been used by both British settlers and indigenous peoples to lay claim to land in the country. Despite apparent land right protections, Maori peoples were systematically disenfranchised by British settlers. In 1975, court cases supported the Treaty and established New Zealand as an officially bilingual country, although Maori experience is still much worse than those of settler ancestry.

4. The Pacific Islands were first integrated into the global economy as European traders became more interested in some of the crops grown and resources found in this region. WW II further brought the Pacific Islands into the global arena as many of these places became sites of military

bases for the Japanese and the United States. During the Cold War, the United States maintained a strong interest in the region militarily. Today, some islands in the region remain dependencies of the United States, France, and Britain. Today, many of the countries are drawn into the global economy because of their dependence on imported goods, transfer payments, infrastructural improvements, and the dependence on tourism for local economies.

5. In 1973, Australia removed its racist immigration policies, replacing racial requirements for skill requirements. This prompted a wave of immigration from Asia, with large numbers of migrants from Vietnam, Hong Kong, and Philippines.

6. Future issues affecting this region include: demographic challenges related to increasing migration to the region as well as migration across the region; challenges of increasing pressure on the wealthier economies of the region to play an even stronger role in future development efforts globally; and the potential for increasing trans-regional cooperation, which may include increasing relations with ASEAN or countries (or blocs) in Latin America, Sub-Saharan Africa, or India (or the Indian Ocean more generally).

7. c	8. b	9. a	10. d
11. a	12. c	13. d	14. a
15. b	16. a	17. F	18. T
19. T	20. F	21. T	22. F
23. T	24. F	25. T	

2.6

The total fertility rate worldwide. TFRs range from 1.7 in more developed countries to 4.6 in the least developed countries.

IN 2009,

world population is 6.8 billion, and by 2025 is projected to increase to 8.1 billion.

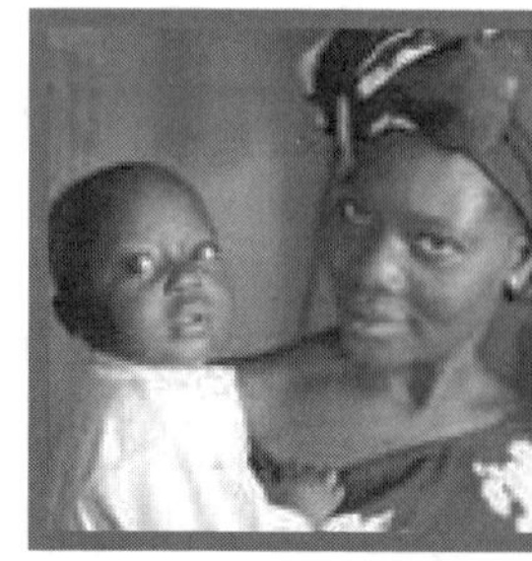

POPULATION REFERENCE BUREAU

2009 WORLD POPULATION DATA SHEET

48%

The share of people worldwide living on less than $2 a day.

BY 2050,

the population of Africa is projected to double to almost 2 billion.

INFORM | EMPOWER | ADVANCE | www.prb.org

MOST POPULOUS COUNTRIES, 2009 AND 2050

2009

COUNTRY	POPULATION (MILLIONS)
China	1,331
India	1,171
United States	307
Indonesia	243
Brazil	191
Pakistan	181
Bangladesh	162
Nigeria	153
Russia	142
Japan	128

2050

COUNTRY	POPULATION (MILLIONS)
India	1,748
China	1,437
United States	439
Indonesia	343
Pakistan	335
Nigeria	285
Bangladesh	222
Brazil	215
Congo, Dem. Rep.	189
Philippines	150

COUNTRIES WITH THE HIGHEST CO2 EMISSIONS, 2006

COUNTRY	METRIC TONS OF CO2 (MILLIONS)
United States	5,697
China	5,607
Russia	1,587
India	1,250
Japan	1,213
Germany	823
Canada	539
United Kingdom	536
South Korea	476
Italy	448

COUNTRY	METRIC TONS PER CAPITA
Qatar	48
Bahrain	27
United Arab Emirates	26
Kuwait	26
Trinidad and Tobago	20
United States	19
Australia	19
Canada	17
Saudi Arabia	14
Finland	13

Note: Countries with a population of less than 1 million excluded.

POPULATION CLOCK, 2009

		WORLD	MORE DEVELOPED COUNTRIES	LESS DEVELOPED COUNTRIES
Population		6,809,972,000	1,232,100,000	5,577,872,000
Births per	Year	138,949,000	14,359,000	124,590,000
	Day	380,683	39,340	341,343
	Minute	264	27	237
Deaths per	Year	56,083,000	12,277,000	43,807,000
	Day	153,653	33,636	120,019
	Minute	107	23	83
Natural increase (births-deaths) per	Year	82,866,000	2,083,000	80,784,000
	Day	227,030	5,707	221,326
	Minute	158	4	154
Infant deaths per	Year	6,352,000	82,000	6,269,000
	Day	17,402	225	17,175
	Minute	12	0.2	12

World Population Growth Is Now Almost Entirely Concentrated in the World's Poorer Countries.

In 2009, world population stood at 6.8 billion, up about 83 million from 2008. The world total is likely to reach 7 billion in the latter half of 2011, with the bulk of growth in the world's poorest nations. The less developed countries of Africa, Asia, and Latin America and the Caribbean are projected to increase by just under 50 percent in the 41 years between now and 2050, and the poorest of these are projected to double in population size over that period. But this scenario assumes that fertility in less developed countries will decline smoothly to the low levels observed in today's more developed countries: about 1.8 children per woman. For fertility to fall to those low levels, many factors are key, including significant increases in the use of family planning in many less developed countries.

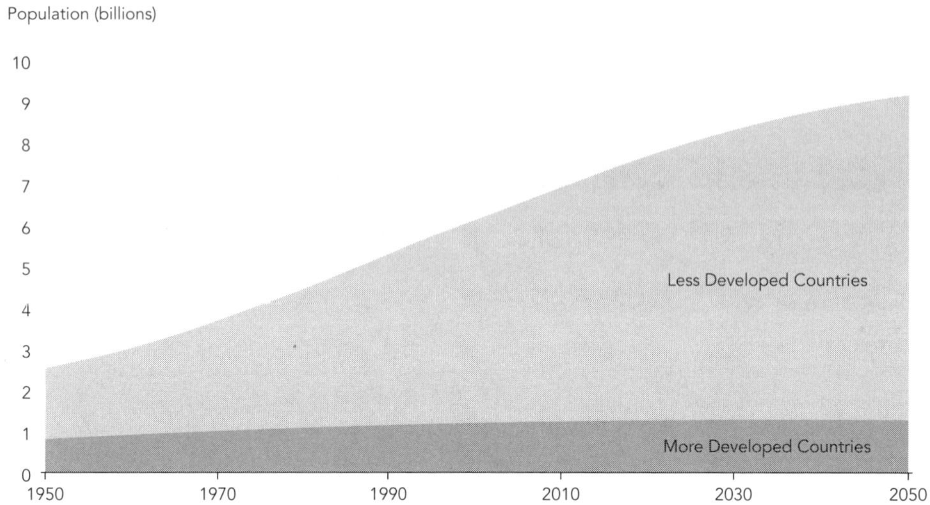

Source: UN Population Division, *World Population Prospects: The 2008 Revision*, medium variant (2009).

The World's Youth Population, Ages 15 to 24, Will Become More and More Concentrated in Africa and Asia.

By 2050, the number of youth will have risen from just under a half billion in 1950 to 1.2 billion. At that point, about nine in 10 youths will be in developing countries. This very large group will arrive at working age with a right to expect gainful employment, adequate health care, and the ability to raise a family with an appropriate living standard if they so choose. Before those things can come about, they must have had access to sufficient education and training so that they can take part in building their country's society and economy. Most likely, tomorrow's youth will have moved to cities in larger numbers as opportunities in the rural areas diminish. But will the economic conditions of their country be able to meet their rising expectations? This will be one of the major social questions of the next few decades.

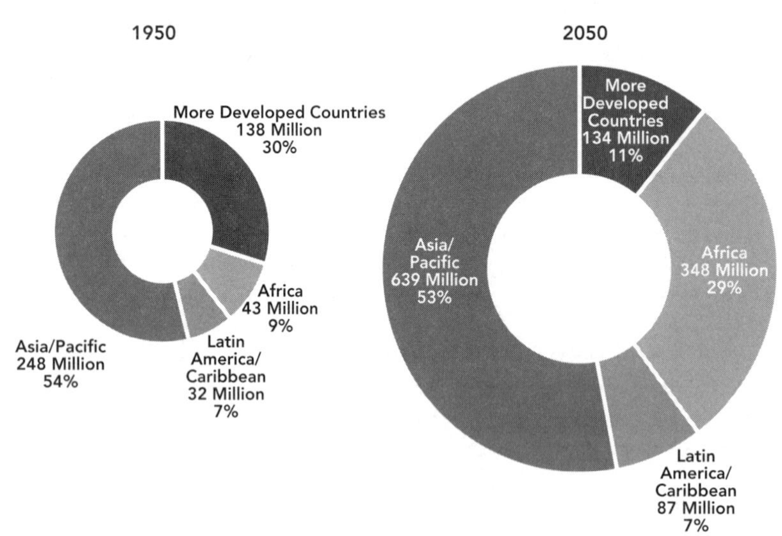

Source: UN Population Division, *World Population Prospects: The 2008 Revision*, medium variant (2009).

The U.S. National Child Poverty Rate Masks Substantial Geographic Differences.

In 2007, 18 percent of American children and youth lived in poverty—but that figure varied widely within the United States. More than 50 percent of the nation's 3,141 counties had child poverty levels exceeding that 18 percent national average. In nearly 1,000 counties at least 25 percent of all children were poor—one in every four children. Counties with the highest child poverty rates were in the Southeast, the Southwest, central Appalachia, much of Alaska, and pockets of the West, Great Plains, and upper Midwest, where rates ranged as high as 67 percent. By contrast, counties with relatively low child poverty were more spread out geographically, with rates as low as 3 percent.

Percent of children under 18 in poverty, 2007

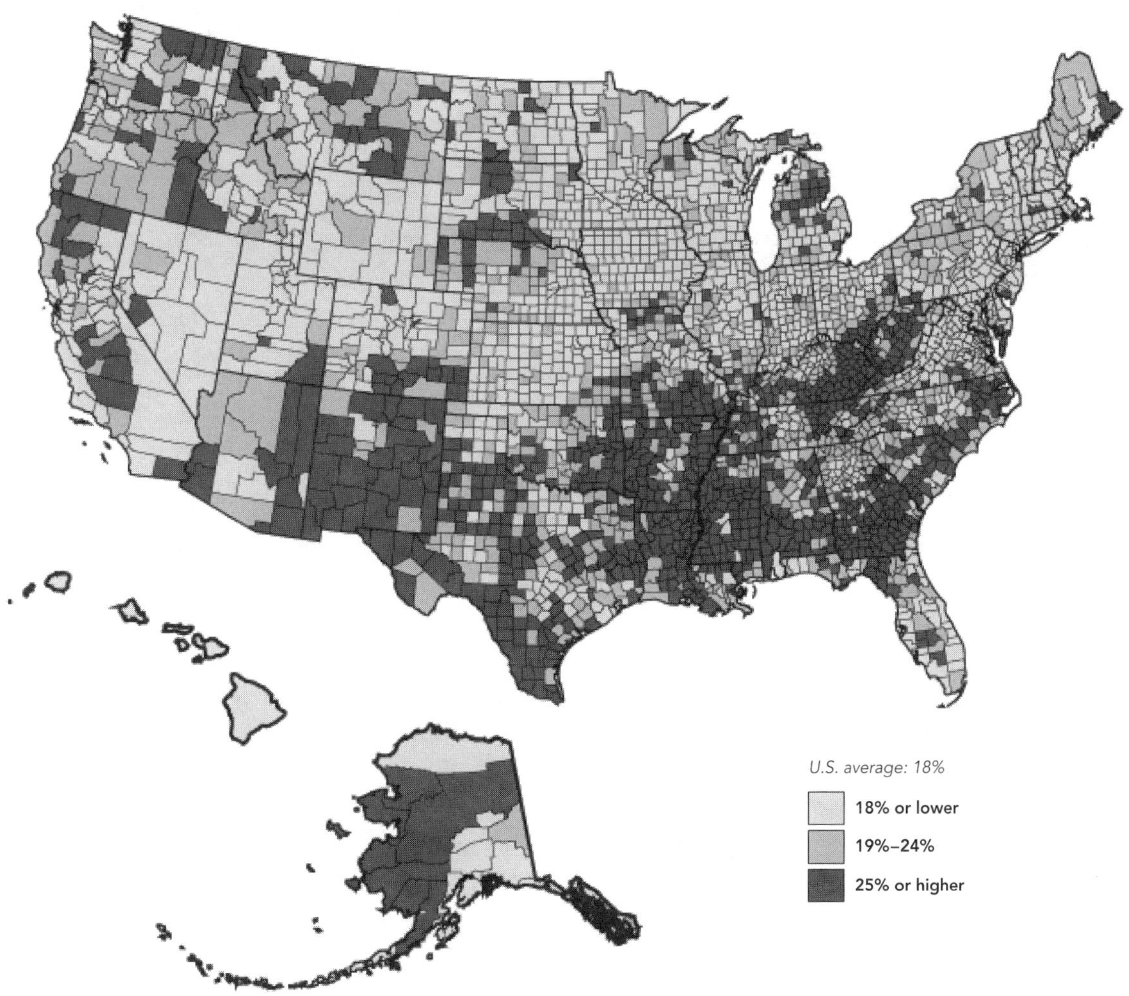

Source: PRB analysis of data from the U.S. Census Bureau, 2007 Small Area Income and Poverty Estimates.

Early Childbearing Poses Serious Health Risks for Both Mother and Child.

Many girls under 18, especially in poorer countries, are physically immature and at higher risk for obstetric complications. And children born to these young mothers are at higher risk for illness and death than children born to mothers in their 20s. Some countries have reduced the prevalence of early childbearing by keeping girls in school and changing community norms and national policies about early marriage.

Percent of females ages 15-19 who are mothers or became pregnant before age 18

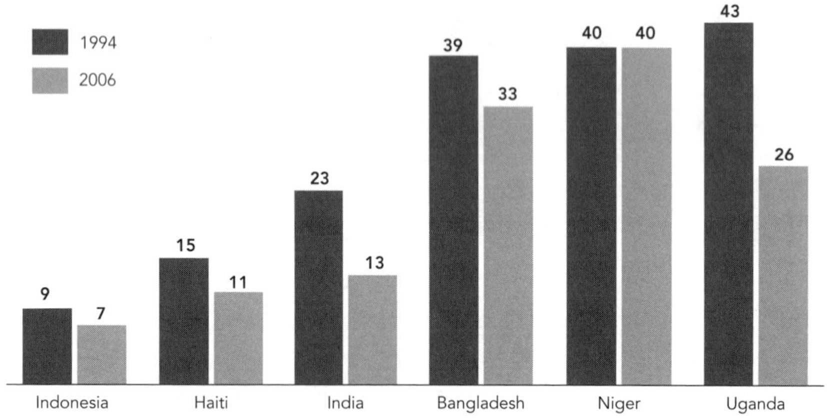

Total fertility rate (lifetime births per woman)

	1994	2006
Indonesia	2.9	2.6
Haiti	4.8	3.9
India	3.5	2.7
Bangladesh	3.4	2.7
Niger	7.0	7.0
Uganda	6.9	6.7

Note: Data are from surveys taken around 1994 and 2006.
Sources: ICF Macro, StatCompiler, accessed online at www.measuredhs.com, on July 20, 2009; and India, Sample Registration System.

Despite Increases, Secondary Enrollment Rates Remain Unacceptably Low in Many Developing Countries.

Education is an obvious necessity for a productive and rewarding life and contributes to economic and social progress for individuals and countries. Children in many developing countries stay in school longer, but those gains have been far from uniform. For many years, a key issue in education has been the lack of equal opportunity for girls but that, too, has undergone notable changes as the ratio of enrolled girls to boys climbs slowly to parity. Often missing in such statistics, however, is information about the quality of education, particularly in rural areas where most students in developing countries live.

Percent enrolled in secondary school, 1990 and 2005

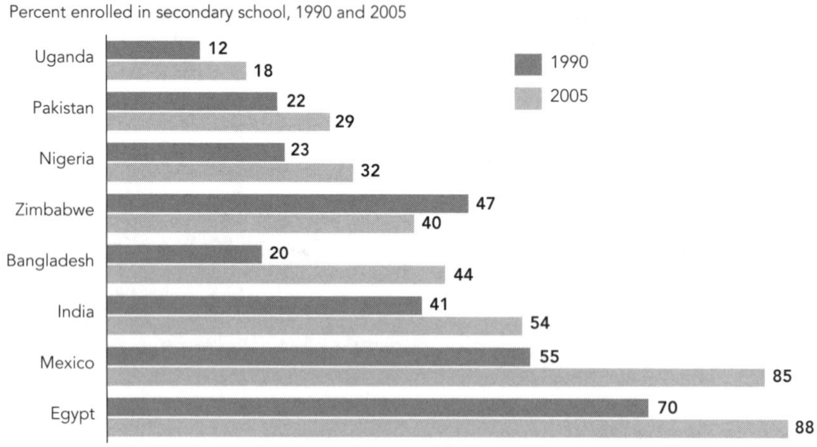

Girls per 100 boys enrolled in secondary school

	1990	2005
Uganda	57	81
Pakistan	42	78
Nigeria	75	82
Zimbabwe	88	93
Bangladesh	52	103
India	58	82
Mexico	98	102
Egypt	77	94

Source: UNESCO Institute for Statistics, Data Centre, accessed online at http://stats.uis.unesco.org, on July 22, 2009.

POPULATION, HEALTH, AND ENVIRONMENT DATA AND ESTIMATES FOR THE COUNTRIES AND REGIONS OF THE WORLD

	Population mid-2009 (millions)	Births per 1,000 Population	Deaths per 1,000 Population	Rate of Natural Increase (%)	Net Migration Rate per 1,000 Population	Projected Population (millions)		Projected Population Change 2009–2050 (%)	Infant Mortality Rate[a]	Total Fertility Rate[b]	Percent of Population Ages	
						mid–2025	mid–2050				<15	65+
WORLD	6,810	20	8	1.2	—	8,087	9,421	38	46	2.6	27	8
MORE DEVELOPED	1,232	12	10	0.2	3	1,282	1,318	7	6	1.7	17	16
LESS DEVELOPED	5,578	22	8	1.4	-1	6,805	8,103	45	50	2.7	30	6
LESS DEVELOPED (Excl. China)	4,246	26	8	1.7	-1	5,329	6,666	57	55	3.1	33	5
LEAST DEVELOPED	828	35	11	2.4	-1	1,151	1,657	100	80	4.6	40	3
AFRICA	999	36	12	2.4	-1	1,385	1,994	100	74	4.8	41	3
SUB-SAHARAN AFRICA	836	39	13	2.5	-1	1,184	1,754	110	80	5.3	43	3
NORTHERN AFRICA	205	25	7	1.9	-1	257	316	54	38	3.0	33	5
Algeria	35.4	23	4	1.9	-1	43.7	50.5	43	26	2.3	28	5
Egypt	78.6	25	6	1.9	-1	99.1	122.3	56	19	3.0	33	5
Libya	6.3	24	4	2.0	1	8.1	9.8	56	18	2.7	30	4
Morocco	31.5	21	6	1.4	-3	36.6	42.4	35	31	2.4	29	6
Sudan	42.3	33	11	2.2	1	56.7	75.9	80	81	4.5	41	3
Tunisia	10.4	17	6	1.2	0	12.2	13.9	34	19	2.0	25	7
Western Sahara	0.5	23	6	1.8	20	0.8	0.9	84	44	3.0	31	2
WESTERN AFRICA	297	40	14	2.7	-1	420	623	110	80	5.5	44	3
Benin	8.9	41	9	3.2	-1	13.8	22.0	146	89	5.7	44	3
Burkina Faso	15.8	46	14	3.2	-1	24.8	40.8	159	89	6.0	46	3
Cape Verde	0.5	26	5	2.1	-5	0.7	0.8	53	29	3.1	38	6
Côte d'Ivoire	21.4	37	14	2.4	-1	29.9	43.7	104	100	4.9	40	2
Gambia	1.6	39	11	2.8	2	2.3	3.6	121	93	5.6	42	3
Ghana	23.8	31	10	2.1	-1	32.2	45.2	90	50	4.0	40	4
Guinea	10.1	39	12	2.7	-6	15.2	24.0	138	104	5.7	43	3
Guinea-Bissau	1.6	43	17	2.6	0	2.3	3.6	121	117	5.9	43	3
Liberia	4.0	40	10	3.0	5	5.9	8.8	124	99	5.8	44	3
Mali	13.0	43	15	2.8	-3	18.6	28.3	117	110	6.0	45	2
Mauritania	3.3	35	10	2.5	-1	4.6	6.9	109	73	5.1	40	4
Niger	15.3	53	14	3.9	0	27.4	58.2	281	88	7.4	49	3
Nigeria	152.6	41	15	2.6	0	207.2	285.1	87	75	5.7	45	3
Senegal	12.5	39	10	2.9	-2	17.9	26.1	108	61	5.0	43	2
Sierra Leone	5.7	40	20	2.0	2	8.1	12.4	119	89	5.2	42	4
Togo	6.6	35	8	2.7	0	9.3	13.2	77	91	5.1	41	3
EASTERN AFRICA	313	40	13	2.6	-1	455	682	118	76	5.4	44	3
Burundi	8.3	36	15	2.1	5	11.2	14.8	79	120	5.4	41	3
Comoros	0.7	33	8	2.5	-3	0.9	1.2	81	53	4.2	38	3
Djibouti	0.9	30	12	1.9	3	1.1	1.5	70	67	4.2	37	3
Eritrea	5.1	38	10	2.9	2	7.4	10.8	113	58	5.3	42	2
Ethiopia	82.8	39	12	2.7	-1	113.1	149.5	81	77	5.3	43	3
Kenya	39.1	39	13	2.7	-1	56.5	83.8	115	67	4.9	42	2
Madagascar	19.5	38	9	2.9	0	28.4	42.3	118	70	5.0	44	3
Malawi	14.2	43	12	3.1	0	21.6	34.1	140	80	6.3	46	3
Mauritius	1.3	14	7	0.7	0	1.4	1.5	16	15.4	1.7	23	7
Mayotte	0.2	39	3	3.6	4	0.3	0.4	105	—	4.5	42	2
Mozambique	22.0	41	17	2.4	0	29.9	42.4	93	97	5.4	43	3
Reunion	0.8	18	5	1.3	0	1.0	1.1	31	8	2.5	27	7
Rwanda	9.9	41	16	2.5	0	14.5	21.8	121	62	5.5	44	3
Seychelles	0.1	18	8	1.0	6	0.1	0.1	37	12.9	2.2	23	8
Somalia	9.1	45	15	3.0	-6	13.9	23.5	158	111	6.7	45	3
Tanzania	43.7	38	15	2.3	-1	67.4	109.5	150	69	5.3	45	3
Uganda	30.7	47	13	3.4	-1	51.8	96.4	214	76	6.7	49	3
Zambia	12.6	45	16	2.9	-1	18.3	28.1	124	70	6.2	46	3
Zimbabwe	12.5	32	18	1.4	-11	16.0	19.1	53	60	3.8	40	4
MIDDLE AFRICA	125	42	14	2.8	0	189	306	144	95	6.1	45	3
Angola	17.1	46	19	2.7	1	26.2	42.7	150	125	6.6	46	2
Cameroon	18.9	36	13	2.3	0	25.5	34.9	85	74	4.7	42	4
Central African Republic	4.5	38	19	1.9	0	5.5	6.5	45	106	5.0	41	4
Chad	10.3	43	17	2.6	-1	13.9	20.5	98	106	6.3	46	3
Congo	3.7	36	13	2.3	-3	5.3	7.8	113	75	5.3	42	3
Congo, Dem. Rep.	68.7	44	13	3.1	0	109.7	189.3	176	92	6.5	47	3
Equatorial Guinea	0.7	38	14	2.4	3	1.0	1.4	114	102	5.4	41	3
Gabon	1.5	28	10	1.8	1	1.9	2.5	68	55	3.6	37	4
Sao Tome and Principe	0.2	34	8	2.6	-2	0.2	0.3	82	75	4.1	41	4

See Notes on page 17.

POPULATION, HEALTH, AND ENVIRONMENT DATA AND ESTIMATES FOR THE COUNTRIES AND REGIONS OF THE WORLD

	Population mid-2009 (millions)	Births per 1,000 Population	Deaths per 1,000 Population	Rate of Natural Increase (%)	Net Migration Rate per 1,000 Population	Projected Population (millions)		Projected Population Change 2009–2050 (%)	Infant Mortality Rate[a]	Total Fertility Rate[b]	Percent of Population Ages	
						mid–2025	mid–2050				<15	65+
SOUTHERN AFRICA	58	24	15	0.9	3	63	68	17	48	2.8	33	5
Botswana	2.0	25	12	1.3	2	2.3	2.8	39	48	3.2	35	5
Lesotho	2.1	25	23	0.2	-4	2.4	2.6	21	83	3.4	39	5
Namibia	2.2	29	8	2.1	0	2.8	3.6	65	46	3.6	38	4
South Africa	50.7	23	15	0.8	3	54.4	57.4	13	45	2.7	32	5
Swaziland	1.2	31	15	1.6	-1	1.5	1.7	48	85	3.8	35	4
AMERICAS	920	17	7	1.1	0	1,063	1,205	31	18	2.2	26	9
NORTHERN AMERICA	341	14	8	0.6	3	395	481	41	6	2.0	20	13
Canada	33.7	11	7	0.4	8	37.6	41.9	24	5.0	1.6	17	14
United States	306.8	14	8	0.6	3	357.5	439.0	43	6.6	2.1	20	13
LATIN AMERICA/CARIBBEAN	580	20	6	1.4	-2	668	724	25	23	2.3	30	6
CENTRAL AMERICA	152	22	5	1.7	-4	179	200	31	21	2.5	33	5
Belize	0.3	27	4	2.3	-1	0.4	0.5	56	18	3.1	39	5
Costa Rica	4.5	16	4	1.3	1	5.3	6.1	35	9.7	1.9	27	6
El Salvador	7.3	21	7	1.4	-3	9.1	11.2	52	24	2.5	34	5
Guatemala	14.0	34	6	2.8	-2	19.9	27.5	96	34	4.4	43	4
Honduras	7.5	27	5	2.2	-1	9.8	12.4	66	23	3.3	38	4
Mexico	109.6	20	5	1.6	-4	123.4	129.0	18	19	2.3	32	6
Nicaragua	5.7	26	5	2.1	-7	7.1	8.2	44	24	2.9	38	4
Panama	3.5	20	4	1.6	1	4.3	5.1	47	15	2.4	30	6
CARIBBEAN	41	19	8	1.2	-3	46	49	19	34	2.5	28	8
Antigua and Barbuda	0.1	17	7	1.0	2	0.1	0.1	26	*20*	2.1	28	7
Bahamas	0.3	17	6	1.1	1	0.4	0.5	33	11	1.9	28	6
Barbados	0.3	13	8	0.5	-1	0.3	0.3	-8	*13*	1.8	22	12
Cuba	11.2	11	8	0.3	-3	11.2	10.0	-11	4.7	1.6	18	12
Dominica	0.1	16	9	0.7	-6	0.1	0.1	-10	*16*	2.3	29	10
Dominican Republic	10.1	24	6	1.8	-3	12.0	13.4	33	32	2.8	33	6
Grenada	0.1	19	7	1.2	-10	0.1	0.1	-8	*14*	2.4	29	6
Guadeloupe	0.4	17	7	1.0	-2	0.5	0.5	21	6	2.4	23	12
Haiti	9.2	31	10	2.1	-2	11.7	15.1	63	57	4.0	38	4
Jamaica	2.7	17	6	1.1	-7	2.8	2.7	-1	26	2.4	30	8
Martinique	0.4	13	7	0.7	-1	0.4	0.4	-13	8	1.9	22	12
Netherlands Antilles	0.2	13	8	0.5	9	0.2	0.2	-7	13	1.9	22	10
Puerto Rico	4.0	12	8	0.4	-1	4.1	3.7	-6	8.8	1.7	20	14
St. Kitts-Nevis	0.05	18	8	1.0	-2	0.1	0.1	26	15	2.3	28	8
Saint Lucia	0.2	15	7	0.8	-1	0.2	0.2	28	19.4	1.7	28	7
St. Vincent & the Grenadines	0.1	17	8	0.9	-9	0.1	0.1	-13	17.6	2.1	29	7
Trinidad and Tobago	1.3	14	8	0.6	-3	1.4	1.3	-4	26	1.6	24	7
SOUTH AMERICA	386	19	6	1.3	-1	443	475	23	23	2.2	29	6
Argentina	40.3	18	8	1.0	0	45.9	50.9	26	12.9	2.4	26	10
Bolivia	9.9	29	8	2.1	-1	12.4	14.9	51	50	3.5	38	4
Brazil	191.5	17	6	1.0	0	212.4	215.3	12	24	2.0	28	6
Chile	17.0	15	5	1.0	0	19.3	20.7	22	7.6	1.9	25	8
Colombia	45.1	20	6	1.4	-1	53.8	59.2	31	19	2.4	30	5
Ecuador	13.6	23	5	1.8	-1	16.1	18.0	32	25	2.8	33	6
French Guiana	0.2	32	4	2.8	6	0.3	0.5	104	10.4	3.9	35	4
Guyana	0.8	21	9	1.2	-10	0.8	0.5	-29	43	2.5	32	5
Paraguay	6.3	27	6	2.1	-1	8.0	9.9	55	36	3.5	36	5
Peru	29.2	21	6	1.5	-4	34.5	39.8	36	26	2.6	32	6
Suriname	0.5	17	7	1.1	-2	0.5	0.5	-8	16	2.4	30	7
Uruguay	3.4	14	9	0.5	-3	3.5	3.6	7	10.5	2.0	24	13
Venezuela	28.4	24	4	2.0	-1	35.1	41.7	47	16.5	2.6	31	5
ASIA	4,117	19	7	1.2	-0	4,858	5,461	33	43	2.3	27	7
ASIA (Excl. China)	2,786	22	7	1.5	-0	3,382	4,024	44	48	2.7	30	6
WESTERN ASIA	231	25	6	1.9	2	293	377	63	38	3.1	33	5
Armenia	3.1	15	10	0.6	-1	3.3	3.3	7	25	1.7	20	10
Azerbaijan	8.8	18	6	1.2	0	9.7	11.6	32	11	2.3	23	7
Bahrain	1.2	15	2	1.3	51	1.6	2.0	61	8	2.0	21	3
Cyprus	1.1	12	7	0.5	9	1.1	1.1	0	6	1.5	18	11
Georgia	4.6	11	9	0.2	-5	4.2	3.5	-23	13	1.4	17	15
Iraq	30.0	32	9	2.3	-4	43.3	61.9	106	84	4.4	41	3

See Notes on page 17.

POPULATION, HEALTH, AND ENVIRONMENT DATA AND ESTIMATES FOR THE COUNTRIES AND REGIONS OF THE WORLD	Population mid-2009 (millions)	Births per 1,000 Population	Deaths per 1,000 Population	Rate of Natural Increase (%)	Net Migration Rate per 1,000 Population	Projected Population (millions) mid-2025	Projected Population (millions) mid-2050	Projected Population Change 2009–2050 (%)	Infant Mortality Rate[a]	Total Fertility Rate[b]	Percent of Population Ages <15	Percent of Population Ages 65+
Israel	7.6	21	5	1.6	1	9.4	11.4	49	3.6	2.9	28	10
Jordan	5.9	28	4	2.4	8	7.6	9.6	62	20	3.6	37	3
Kuwait	3.0	18	2	1.6	8	4.0	5.2	76	9	2.2	23	2
Lebanon	3.9	20	5	1.5	-1	4.5	5.3	37	19	2.4	26	7
Oman	3.1	25	3	2.2	32	4.1	5.3	71	10	3.4	29	2
Palestinian Territory	3.9	33	4	2.8	-1	6.0	9.4	140	25	4.6	44	3
Qatar	1.4	13	1	1.1	94	1.8	2.3	64	7	2.4	17	1
Saudi Arabia	28.7	29	2	2.6	-7	35.7	49.8	74	18	3.9	38	2
Syria	21.9	28	3	2.5	8	28.6	36.9	68	16	3.3	36	3
Turkey	74.8	18	6	1.2	0	87.4	97.4	30	28	2.1	27	6
United Arab Emirates	5.1	15	2	1.4	48	6.7	9.1	79	7	2.0	19	1
Yemen	22.9	38	8	3.0	-1	34.5	52.1	128	69	5.5	45	3
SOUTH CENTRAL ASIA	1,726	24	7	1.7	-0	2,148	2,624	52	57	2.8	32	5
Afghanistan	28.4	39	18	2.1	6	39.4	53.4	88	155	5.7	44	2
Bangladesh	162.2	23	7	1.6	-1	195.0	222.5	37	48	2.5	32	4
Bhutan	0.7	25	8	1.7	3	0.9	1.0	46	40	3.1	32	5
India	1,171.0	23	7	1.6	-0	1,444.5	1,748.0	49	55	2.7	32	5
Iran	73.2	20	5	1.5	-1	88.0	100.2	37	35	2.0	28	5
Kazakhstan	15.9	23	10	1.3	0	18.0	21.1	33	32	2.7	24	8
Kyrgyzstan	5.3	24	7	1.6	-10	6.5	8.0	52	31	2.8	30	5
Maldives	0.3	22	4	1.8	0	0.4	0.5	51	10	2.3	30	5
Nepal	27.5	29	9	2.1	-1	35.7	46.0	67	48	3.1	37	4
Pakistan	180.8	30	7	2.3	-2	246.3	335.2	85	67	4.0	38	4
Sri Lanka	20.5	19	7	1.2	-2	23.2	25.4	24	15	2.4	26	7
Tajikistan	7.5	28	5	2.3	-2	10.2	13.7	84	65	3.4	38	4
Turkmenistan	5.1	22	8	1.4	-1	6.1	6.8	33	51	2.5	31	4
Uzbekistan	27.6	23	5	1.8	-2	34.4	42.4	54	48	2.6	33	5
SOUTHEAST ASIA	597	21	7	1.4	-0	712	827	39	30	2.5	29	6
Brunei	0.4	16	3	1.3	2	0.5	0.6	64	7	1.7	26	4
Cambodia	14.8	25	8	1.7	-0	19.0	23.8	61	62	3.0	35	3
Indonesia	243.3	21	6	1.5	-1	291.9	343.1	41	34	2.5	29	6
Laos	6.3	28	7	2.1	-2	8.3	10.7	70	64	3.5	39	4
Malaysia	28.3	21	5	1.6	1	34.6	40.4	43	9	2.6	32	4
Myanmar	50.0	21	10	1.1	-2	57.6	63.4	27	75	2.3	27	5
Philippines	92.2	26	5	2.1	-2	120.2	150.1	63	23	3.3	35	4
Singapore	5.1	10	4	0.6	49	5.8	5.6	10	2.1	1.3	18	9
Thailand	67.8	15	9	0.6	1	72.6	73.4	8	7	1.8	22	7
Timor-Leste	1.1	40	9	3.1	2	1.9	3.2	184	67	6.5	45	3
Vietnam	87.3	17	5	1.2	0	100.1	112.8	29	15	2.1	26	7
EAST ASIA	1,564	12	7	0.5	0	1,704	1,633	4	20	1.6	19	10
China	1,331.4	12	7	0.5	0	1,476.0	1,437.0	8	21	1.6	19	8
China, Hong Kong SAR[c]	7.0	11	6	0.5	3	8.0	8.8	25	1.8	1.1	13	13
China, Macao SAR[c]	0.6	9	3	0.5	15	0.8	0.8	53	*3*	1.0	13	7
Japan	127.6	9	9	-0.0	-0	119.3	95.2	-25	2.6	1.4	13	23
Korea, North	22.7	16	10	0.5	0	23.6	22.9	1	54	2.0	22	9
Korea, South	48.7	9	5	0.4	2	49.1	42.3	-13	4	1.2	17	10
Mongolia	2.7	24	6	1.8	-0	3.3	4.0	48	41	2.6	33	4
Taiwan	23.1	8	6	0.2	1	23.9	21.5	-7	4.7	1.0	17	10
EUROPE	738	11	11	-0.0	3	736	702	-5	6	1.5	15	16
NORTHERN EUROPE	99	13	10	0.3	4	109	118	20	4	1.9	18	16
Channel Islands	0.2	11	8	0.3	6	0.2	0.2	-4	*3.6*	1.4	16	15
Denmark	5.5	12	10	0.2	5	5.7	5.8	5	4.0	1.9	19	17
Estonia	1.3	12	12	-0.0	-0	1.3	1.2	-12	4.4	1.7	15	17
Finland	5.3	11	9	0.2	3	5.6	5.7	8	2.6	1.9	17	17
Iceland	0.3	15	6	0.9	3	0.4	0.4	27	*2.5*	2.1	21	12
Ireland	4.5	16	6	1.0	9	5.7	6.4	42	3.1	2.0	21	11
Latvia	2.3	10	14	-0.3	-1	2.1	1.8	-20	8.7	1.4	14	17
Lithuania	3.3	11	13	-0.2	-3	3.2	2.7	-18	4.9	1.5	15	16
Norway	4.8	13	9	0.4	8	5.6	6.6	36	2.7	2.0	19	15
Sweden	9.3	12	10	0.2	6	9.9	10.5	13	2.5	1.9	17	18
United Kingdom	61.8	13	9	0.4	3	68.8	76.9	24	4.6	1.9	18	16

See Notes on page 17.

POPULATION, HEALTH, AND ENVIRONMENT DATA AND ESTIMATES FOR THE COUNTRIES AND REGIONS OF THE WORLD

	Population mid-2009 (millions)	Births per 1,000 Population	Deaths per 1,000 Population	Rate of Natural Increase (%)	Net Migration Rate per 1,000 Population	Projected Population (millions) mid–2025	Projected Population (millions) mid–2050	Projected Population Change 2009–2050 (%)	Infant Mortality Rate[a]	Total Fertility Rate[b]	Percent of Population Ages <15	Percent of Population Ages 65+
WESTERN EUROPE	189	10	9	0.1	1	192	189	0	4	1.6	16	18
Austria	8.4	9	9	0.0	4	8.8	9.5	14	3.7	1.4	15	17
Belgium	10.8	11	9	0.2	4	11.8	12.5	15	4.0	1.7	17	17
France	62.6	13	9	0.4	1	66.1	70.0	12	3.6	2.0	18	17
Germany	82.0	8	10	-0.2	0	79.6	71.4	-13	3.9	1.3	14	20
Liechtenstein	0.04	10	6	0.4	5	0.04	0.05	26	*2.7*	1.4	17	12
Luxembourg	0.5	11	7	0.4	16	0.6	0.7	40	1.8	1.6	18	14
Monaco	0.04	7	6	0.0	0	0.04	0.04	7	—	—	13	24
Netherlands	16.5	11	8	0.3	2	17.2	17.3	5	3.8	1.8	18	15
Switzerland	7.8	10	8	0.2	7	8.1	8.1	4	3.9	1.5	15	17
EASTERN EUROPE	295	11	14	-0.2	1	278	243	-18	8	1.5	15	14
Belarus	9.7	11	14	-0.3	1	9.2	8.4	-13	5	1.4	15	14
Bulgaria	7.6	10	14	-0.4	-0	6.9	5.9	-22	8.6	1.5	13	17
Czech Republic	10.5	11	10	0.1	7	10.2	9.4	-10	2.8	1.5	14	15
Hungary	10.0	10	13	-0.3	2	9.8	9.1	-10	5.3	1.3	15	16
Moldova	4.1	11	12	-0.1	-1	3.8	3.1	-24	12	1.3	18	10
Poland	38.1	11	10	0.1	-0	37.4	31.8	-17	5.6	1.4	15	14
Romania	21.5	10	12	-0.1	-0	20.5	18.1	-15	10.9	1.3	15	15
Russia	141.8	12	15	-0.3	2	133.3	116.9	-18	9	1.5	15	14
Slovakia	5.4	11	10	0.1	1	5.4	4.9	-10	5.4	1.3	15	12
Ukraine	46.0	11	16	-0.5	0	41.9	35.3	-23	10	1.4	14	16
SOUTHERN EUROPE	155	10	9	0.1	6	157	151	-3	5	1.4	15	18
Albania	3.2	10	5	0.6	-0	3.3	2.9	-11	6	1.3	25	9
Andorra	0.1	10	3	0.7	16	0.1	0.1	-5	*3.6*	1.2	15	12
Bosnia-Herzegovina	3.8	9	9	0.0	-0	3.7	3.1	-20	5	1.2	16	14
Croatia	4.4	10	12	-0.2	2	4.3	3.8	-13	5.6	1.5	15	17
Greece	11.3	10	10	0.0	4	11.7	11.5	2	3.6	1.4	14	19
Italy	60.3	10	10	-0.0	8	61.9	61.7	2	3.7	1.4	14	20
Kosovo[d]	2.2	21	7	1.4	-	2.7	3.2	43	33	2.5	31	6
Macedonia[e]	2.0	11	9	0.2	0	2.0	1.9	-9	10	1.5	19	11
Malta	0.4	9	8	0.2	5	0.4	0.4	-3	*6.5*	1.4	16	14
Montenegro	0.6	13	10	0.3	-1	0.6	0.6	-4	*7.4*	1.7	20	13
Portugal	10.6	10	10	-0.0	2	10.9	10.7	0	3.4	1.3	15	17
San Marino	0.03	10	7	0.3	12	0.04	0.04	11	—	1.2	15	16
Serbia	7.3	9	14	-0.5	1	6.8	5.9	-20	7.1	1.4	15	17
Slovenia	2.0	11	9	0.2	9	2.0	1.9	-8	*2.1*	1.5	14	16
Spain	46.9	11	8	0.3	8	46.2	43.9	-7	3.5	1.5	14	17
OCEANIA	36	18	7	1.1	7	45	58	60	22	2.5	24	10
Australia	21.9	14	7	0.7	12	26.9	34.0	55	4.1	2.0	19	13
Federated States of Micronesia	0.1	25	6	1.9	-15	0.1	0.1	24	38	3.9	37	4
Fiji	0.8	24	7	1.7	-8	0.9	0.9	12	18	2.6	29	5
French Polynesia	0.3	18	4	1.3	0	0.3	0.4	34	*6.8*	2.2	26	6
Guam	0.2	19	4	1.5	0	0.2	0.3	47	*13.3*	2.6	28	7
Kiribati	0.1	27	9	1.8	-1	0.1	0.2	65	52	3.5	36	4
Marshall Islands	0.1	35	6	2.9	-23	0.1	0.1	13	*21*	4.5	41	2
Nauru	0.01	31	10	2.1	0	0.01	0.02	64	46	3.4	39	1
New Caledonia	0.3	17	5	1.2	5	0.3	0.4	43	*6*	2.2	27	7
New Zealand	4.3	15	7	0.8	2	4.9	5.5	27	5.0	2.2	21	13
Palau	0.02	13	7	0.6	2	0.02	0.02	12	*20*	2.0	24	6
Papua New Guinea	6.6	31	10	2.2	0	9.1	13.4	103	51	4.1	40	2
Samoa	0.2	26	5	2.0	-17	0.2	0.2	11	20	4.2	40	5
Solomon Islands	0.5	35	8	2.7	-3	0.8	1.2	140	48	4.6	41	3
Tonga	0.1	27	7	2.1	-17	0.1	0.1	19	*19*	4.2	38	6
Tuvalu	0.01	23	9	1.4	-9	0.01	0.01	26	35	3.7	32	6
Vanuatu	0.2	31	6	2.5	0	0.3	0.5	126	*27*	4.0	41	3

See Notes on page 17.

POPULATION, HEALTH, AND ENVIRONMENT DATA AND ESTIMATES FOR THE COUNTRIES AND REGIONS OF THE WORLD

	Life Expectancy at Birth (years)			Percent Urban	Percent of Population With HIV/AIDS				Percent of Married Women 15–49 Using Contraception*	
					Adults Ages 15–49		Youth Ages 15–24, 2007/2008			
	Both Sexes	Males	Females		2001	2007/2008	Males	Females	All Methods	Modern Methods
WORLD	69	67	71	50	0.8	0.8	0.4	0.6	62	55
MORE DEVELOPED	77	74	81	75	0.4	0.5	0.6	0.3	68	58
LESS DEVELOPED	67	65	68	44	1.0	1.0	0.4	0.8	61	55
LESS DEVELOPED (Excl. China)	65	63	67	43	1.4	1.4	0.6	1.0	51	43
LEAST DEVELOPED	56	55	57	27	—	2.9	0.9	1.9	28	22
AFRICA	55	53	56	38	4.6	4.3	1.1	3.0	28	23
SUB-SAHARAN AFRICA	51	50	53	35	5.7	5.0	1.2	3.4	22	17
NORTHERN AFRICA	69	67	71	50	0.4	0.4	—	—	49	44
Algeria	72	71	74	63	0.1	0.1	0.1	0.1	61	52
Egypt	72	70	74	43	—	—	—	—	60	58
Libya	73	71	76	77	—	—	—	—	42	20
Morocco	71	69	73	56	0.1	0.1	0.1	0.1	63	55
Sudan	58	57	60	38	1.4	1.4	0.3	1.0	8	6
Tunisia	74	72	76	66	<0.1	0.1	0.1	<0.1	60	52
Western Sahara	65	63	67	81	—	—	—	—	—	—
WESTERN AFRICA	51	50	52	42	2.8	2.5	0.7	1.8	15	9
Benin	56	54	57	41	1.3	1.2	0.3	0.9	17	6
Burkina Faso	57	56	58	16	2.1	1.6	0.5	0.9	17	13
Cape Verde	71	68	73	59	—	—	—	—	61	57
Côte d'Ivoire	52	50	53	48	6.0	3.9	0.8	2.4	13	8
Gambia	55	54	57	54	0.9	0.9	0.2	0.6	10	9
Ghana	59	58	59	48	2.3	1.9	0.4	1.3	24	17
Guinea	56	54	58	33	1.2	1.6	0.4	1.2	9	6
Guinea-Bissau	46	45	48	30	1.8	1.8	0.4	1.2	10	6
Liberia	56	54	57	58	1.4	1.7	0.4	1.3	11	10
Mali	48	47	48	31	1.5	1.5	0.4	1.1	8	6
Mauritania	57	55	59	40	0.7	0.8	0.9	0.5	9	8
Niger	53	51	54	17	0.7	0.8	0.9	0.5	11	5
Nigeria	47	47	48	47	3.2	3.1	0.8	2.3	15	9
Senegal	55	54	57	41	0.4	1.0	0.3	0.8	12	10
Sierra Leone	48	48	49	37	—	1.5	0.6	1.4	8	7
Togo	61	60	63	40	3.6	3.3	0.8	2.4	17	11
EASTERN AFRICA	51	50	52	22	—	5.6	1.2	3.6	26	21
Burundi	49	48	51	10	3.5	2.0	0.4	1.3	9	8
Comoros	64	62	66	28	<0.1	<0.1	0.1	<0.1	*26*	*19*
Djibouti	55	53	56	87	3.1	3.1	0.7	2.1	18	17
Eritrea	58	56	61	21	1.2	1.3	0.3	0.9	*8*	*5*
Ethiopia	53	51	54	16	2.4	2.1	0.5	1.5	15	14
Kenya	54	53	55	19	—	7.4	1.4	5.6	39	32
Madagascar	59	57	61	30	0.1	0.1	0.2	0.1	27	17
Malawi	46	45	47	17	13.3	11.9	2.4	8.4	42	39
Mauritius	72	69	76	42	0.3	1.7	1.8	1.0	*76*	42
Mayotte	74	72	76	28	—	—	—	—	—	—
Mozambique	43	42	44	29	10.3	12.5	2.9	8.5	17	12
Reunion	76	72	80	92	—	—	—	—	*67*	*64*
Rwanda	48	46	50	18	4.3	2.8	0.5	1.4	36	27
Seychelles	73	68	79	53	—	—	—	—	—	—
Somalia	50	48	51	37	0.5	0.5	0.6	0.3	15	1
Tanzania	54	53	55	25	—	5.7	0.5	0.9	26	20
Uganda	50	50	51	13	7.9	5.4	1.3	3.9	24	18
Zambia	43	43	44	37	—	14.3	4.3	8.5	41	33
Zimbabwe	41	39	43	37	26.0	15.3	2.9	7.7	60	58
MIDDLE AFRICA	51	49	52	41	—	2.5	1.0	1.6	19	7
Angola	46	44	48	57	1.6	2.1	0.2	0.3	6	5
Cameroon	52	51	52	57	6.0	5.1	1.2	4.3	26	12
Central African Republic	45	45	45	38	6.4	6.3	1.1	5.5	19	9
Chad	47	46	48	27	3.4	3.5	2.0	2.8	3	2
Congo	53	52	54	60	4.4	3.5	0.8	2.3	44	13
Congo, Dem. Rep.	53	49	55	33	—	1.3	1.0	0.5	21	6
Equatorial Guinea	59	59	60	39	3.7	3.4	0.8	2.5	—	—
Gabon	59	58	61	84	5.6	5.9	1.3	3.9	*33*	*12*
Sao Tome and Principe	65	63	67	58	—	—	—	—	*29*	*27*

See Notes on page 17.

POPULATION, HEALTH, AND ENVIRONMENT DATA AND ESTIMATES FOR THE COUNTRIES AND REGIONS OF THE WORLD

	Life Expectancy at Birth (years)			Percent Urban	Percent of Population With HIV/AIDS: Adults Ages 15–49		Youth Ages 15–24, 2007/2008		Percent of Married Women 15–49 Using Contraception*	
	Both Sexes	Males	Females		2001	2007/2008	Males	Females	All Methods	Modern Methods
SOUTHERN AFRICA	52	50	53	56	17.6	18.5	4.1	13.0	59	58
Botswana	49	44	54	60	26.5	23.9	5.1	15.3	*44*	*42*
Lesotho	40	40	39	24	23.9	23.2	5.9	14.9	37	35
Namibia	59	58	61	35	14.6	15.3	3.4	10.3	55	53
South Africa	52	50	54	59	16.9	18.1	4.0	12.7	60	60
Swaziland	46	46	46	24	26.3	26.1	5.8	22.6	51	48
AMERICAS	75	72	78	78	0.5	0.6	0.7	0.4	72	66
NORTHERN AMERICA	78	75	80	79	0.6	0.6	0.7	0.3	73	69
Canada	81	78	83	81	0.3	0.4	0.4	0.2	*75*	*73*
United States	78	75	80	79	0.6	0.6	0.7	0.3	*73*	*68*
LATIN AMERICA/CARIBBEAN	73	70	76	77	0.5	0.5	0.7	0.4	71	64
CENTRAL AMERICA	74	72	77	70	0.4	0.4	0.4	0.4	68	63
Belize	73	71	74	51	2.1	2.1	0.5	1.5	*56*	*49*
Costa Rica	79	77	82	59	0.2	0.4	0.4	0.2	*80*	*72*
El Salvador	71	66	76	60	0.8	0.8	0.9	0.5	73	66
Guatemala	70	66	73	47	0.8	0.8	—	1.5	*43*	*34*
Honduras	72	69	74	49	0.9	0.7	0.7	0.4	65	56
Mexico	75	73	78	77	0.3	0.3	0.3	0.2	*71*	*67*
Nicaragua	71	68	74	58	0.2	0.2	0.3	0.1	72	70
Panama	75	73	78	64	1.0	1.0	1.1	0.6	—	—
CARIBBEAN	71	69	74	63	1.1	1.1	0.5	0.8	62	55
Antigua and Barbuda	73	71	75	31	—	—	—	—	—	—
Bahamas	72	70	75	83	3.1	3.0	3.2	1.5	—	—
Barbados	77	73	79	38	1.2	1.2	1.3	0.6	—	—
Cuba	78	76	80	76	<0.1	0.1	0.1	0.1	73	72
Dominica	75	72	77	73	—	—	—	—	—	—
Dominican Republic	72	69	75	64	1.3	1.1	0.3	0.6	73	60
Grenada	74	73	76	31	—	—	—	—	—	—
Guadeloupe	79	75	83	100	—	—	—	—	—	—
Haiti	58	57	60	43	2.2	2.2	0.6	1.4	32	25
Jamaica	72	68	75	52	1.4	1.6	1.7	0.9	*69*	*66*
Martinique	80	76	83	98	—	—	—	—	—	—
Netherlands Antilles	76	71	79	92	—	—	—	—	—	—
Puerto Rico	78	74	82	94	—	—	—	—	*78*	*68*
St. Kitts-Nevis	70	68	72	32	—	—	—	—	—	—
Saint Lucia	73	71	76	28	—	—	—	—	—	—
St. Vincent & the Grenadines	72	70	74	40	—	—	—	—	—	—
Trinidad and Tobago	69	67	71	12	1.4	1.5	0.3	1.0	43	38
SOUTH AMERICA	73	70	76	81	0.5	0.6	0.8	0.5	73	65
Argentina	75	71	79	91	0.5	0.5	0.6	0.3	*65*	—
Bolivia	65	63	67	65	0.1	0.2	0.2	0.1	61	35
Brazil	73	69	77	84	0.6	0.6	1.0	0.6	*76*	*70*
Chile	78	75	81	87	0.3	0.3	0.3	0.2	*61*	*58*
Colombia	72	69	76	75	0.5	0.6	0.7	0.3	78	68
Ecuador	75	72	78	63	0.3	0.3	0.4	0.2	73	59
French Guiana	75	72	79	76	—	—	—	—	—	—
Guyana	66	63	69	28	2.5	2.5	0.5	1.7	34	33
Paraguay	71	69	73	57	0.4	0.6	0.7	0.3	79	71
Peru	72	70	75	76	0.4	0.5	0.5	0.3	71	48
Suriname	69	66	73	67	1.3	2.4	2.7	1.4	42	41
Uruguay	76	72	79	94	0.4	0.6	0.6	0.3	77	75
Venezuela	73	70	76	88	—	—	—	—	*70*	*62*
ASIA	69	68	71	42	0.3	0.2	0.2	0.2	67	61
ASIA (Excl. China)	68	66	69	40	—	0.3	0.3	0.3	56	47
WESTERN ASIA	72	69	74	64	—	—	—	—	54	35
Armenia	72	68	75	64	0.1	0.1	0.2	0.1	53	20
Azerbaijan	72	70	75	52	—	0.2	0.3	0.1	51	14
Bahrain	75	73	77	100	—	—	—	—	—	—
Cyprus	78	75	80	62	—	—	—	—	—	—
Georgia	75	71	79	53	—	0.1	0.1	0.1	47	27
Iraq	67	64	72	67	—	—	—	—	50	33

See Notes on page 17.

POPULATION, HEALTH, AND ENVIRONMENT DATA AND ESTIMATES FOR THE COUNTRIES AND REGIONS OF THE WORLD

	Life Expectancy at Birth (years)			Percent Urban	Percent of Population With HIV/AIDS: Adults Ages 15–49		Percent of Population With HIV/AIDS: Youth Ages 15–24, 2007/2008		Percent of Married Women 15–49 Using Contraception*	
	Both Sexes	Males	Females		2001	2007/2008	Males	Females	All Methods	Modern Methods
Israel	81	79	83	92	0.1	0.1	<0.1	0.1	—	—
Jordan	73	72	74	83	—	—	—	—	57	42
Kuwait	78	76	80	98	—	—	—	—	*52*	*39*
Lebanon	72	70	74	87	0.1	0.1	0.1	0.1	58	34
Oman	72	70	74	71	—	—	—	—	—	—
Palestinian Territory	72	70	73	72	—	—	—	—	50	39
Qatar	76	75	77	100	—	—	—	—	*43*	*32*
Saudi Arabia	76	74	78	81	—	—	—	—	*32*	*29*
Syria	74	72	76	54	—	—	—	—	58	43
Turkey	72	69	74	63	—	—	—	—	71	43
United Arab Emirates	77	77	79	83	—	—	—	—	—	—
Yemen	63	62	64	29	—	—	—	—	28	19
SOUTH CENTRAL ASIA	65	64	66	31	0.4	0.3	0.3	0.3	53	45
Afghanistan	44	44	44	22	—	—	—	—	10	9
Bangladesh	65	64	66	25	—	—	—	—	56	48
Bhutan	68	67	68	31	—	0.1	0.1	<0.1	—	*31*
India	64	63	65	29	0.5	0.3	0.3	0.3	56	49
Iran	71	69	73	67	0.1	0.2	0.2	0.1	*74*	*56*
Kazakhstan	67	62	72	53	<0.1	0.1	0.2	0.1	51	49
Kyrgyzstan	68	64	72	35	<0.1	0.1	0.2	0.1	48	46
Maldives	73	72	74	35	—	—	—	—	39	34
Nepal	64	63	64	17	0.5	0.5	0.5	0.3	48	44
Pakistan	66	66	67	35	<0.1	0.1	0.1	0.1	30	22
Sri Lanka	71	67	75	15	—	—	<0.1	—	68	53
Tajikistan	67	64	69	26	0.1	0.3	0.4	0.1	38	33
Turkmenistan	65	61	69	47	—	<0.1	—	—	*62*	*53*
Uzbekistan	68	65	71	36	—	0.1	0.1	0.1	65	59
SOUTHEAST ASIA	70	68	72	43	0.5	0.5	0.5	0.3	60	53
Brunei	77	75	80	72	—	—	—	—	—	—
Cambodia	61	59	63	15	1.5	0.8	0.8	0.3	40	27
Indonesia	71	69	73	43	0.1	0.2	0.3	0.1	61	57
Laos	65	63	66	27	<0.1	0.2	0.2	0.1	*32*	*29*
Malaysia	74	72	77	68	0.3	0.5	0.6	0.3	—	—
Myanmar	61	59	63	31	0.9	0.7	0.7	0.6	*37*	*33*
Philippines	69	66	72	63	—	—	—	—	51	34
Singapore	81	78	83	100	0.1	0.2	0.2	0.1	*62*	*55*
Thailand	69	66	72	36	1.7	1.4	1.2	1.2	72	70
Timor-Leste	61	60	62	22	—	—	—	—	10	9
Vietnam	74	72	76	28	0.3	0.5	0.6	0.3	79	68
EAST ASIA	74	72	76	51	0.1	0.1	0.1	0.1	86	85
China	73	71	75	46	0.1	0.1	0.1	0.1	90	90
China, Hong Kong SAR[c]	82	79	86	100	—	—	—	—	—	—
China, Macao SAR[c]	82	79	84	100	—	—	—	—	—	—
Japan	83	79	86	86	—	—	—	—	52	44
Korea, North	63	61	66	60	—	—	—	—	*69*	*58*
Korea, South	80	76	83	82	<0.1	<0.1	<0.1	<0.1	*81*	*67*
Mongolia	65	62	69	60	—	0.1	0.1	—	66	61
Taiwan	78	75	82	78	—	—	—	—	*71*	—
EUROPE	76	72	80	71	0.3	0.5	0.6	0.4	68	55
NORTHERN EUROPE	79	77	81	77	0.2	0.2	0.3	0.1	81	75
Channel Islands	79	77	82	31	—	—	—	—	—	—
Denmark	79	76	81	72	0.1	0.2	0.2	0.1	—	—
Estonia	73	67	79	69	0.5	1.3	1.6	0.7	*70*	*56*
Finland	80	76	83	63	0.1	0.1	0.1	<0.1	*79*	*78*
Iceland	81	80	83	93	0.2	0.2	0.2	0.1	—	—
Ireland	79	77	82	60	0.2	0.2	0.2	0.1	—	—
Latvia	72	66	77	68	0.3	0.8	0.9	0.5	*85*	*60*
Lithuania	71	65	77	67	0.1	0.1	0.1	0.1	*47*	*30*
Norway	81	78	83	80	0.1	0.1	0.1	0.1	—	—
Sweden	81	79	83	84	0.1	0.1	0.1	0.1	*75*	*65*
United Kingdom	79	77	82	80	0.2	0.2	0.3	0.1	84	79

POPULATION, HEALTH, AND ENVIRONMENT DATA AND ESTIMATES FOR THE COUNTRIES AND REGIONS OF THE WORLD

	Life Expectancy at Birth (years)			Percent Urban	Percent of Population With HIV/AIDS				Percent of Married Women 15–49 Using Contraception*	
					Adults Ages 15–49		Youth Ages 15–24, 2007/2008			
	Both Sexes	Males	Females		2001	2007/2008	Males	Females	All Methods	Modern Methods
WESTERN EUROPE	80	77	83	75	0.2	0.2	0.2	0.1	74	69
Austria	80	78	83	67	0.1	0.2	0.2	0.1	57	41
Belgium	80	77	82	97	0.2	0.2	0.2	0.1	78	74
France	81	78	84	77	0.4	0.4	0.4	0.2	79	76
Germany	80	77	82	73	0.1	0.1	0.1	0.1	70	66
Liechtenstein	80	79	82	15	—	—	—	—	—	—
Luxembourg	80	78	83	83	0.2	0.2	0.2	0.1	—	—
Monaco	—	—	—	100	—	—	—	—	—	—
Netherlands	80	78	82	66	0.2	0.2	0.2	0.1	—	69
Switzerland	82	79	84	73	0.6	0.6	0.4	0.5	82	78
EASTERN EUROPE	70	65	75	69	0.4	0.9	1.0	0.6	64	44
Belarus	70	65	76	74	0.2	0.2	0.3	0.1	73	57
Bulgaria	73	70	77	71	—	—	—	—	42	26
Czech Republic	77	74	80	74	—	—	<0.1	—	72	63
Hungary	73	69	77	66	0.1	0.1	0.1	<0.1	77	68
Moldova	69	65	73	41	<0.1	0.4	0.4	0.2	68	44
Poland	76	71	80	61	0.1	0.1	0.1	0.1	49	19
Romania	72	69	76	55	0.1	0.1	0.2	0.2	70	38
Russia	68	61	74	73	0.5	1.1	1.3	0.6	65	47
Slovakia	75	71	79	56	—	<0.1	—	—	74	41
Ukraine	68	63	74	68	0.8	1.6	1.5	1.5	67	48
SOUTHERN EUROPE	80	77	83	67	0.4	0.4	—	—	62	46
Albania	75	72	79	49	—	—	—	—	60	22
Andorra	—	—	—	90	—	—	—	—	—	—
Bosnia-Herzegovina	75	72	77	46	—	<0.1	—	—	36	11
Croatia	76	72	79	56	—	<0.1	—	—	—	—
Greece	80	77	82	60	0.1	0.2	0.2	0.1	61	34
Italy	82	79	84	68	0.4	0.4	0.4	0.2	60	39
Kosovo[d]	69	67	71	—	—	—	—	—	44	18
Macedonia[e]	74	71	76	65	—	<0.1	—	—	14	10
Malta	80	77	82	94	0.1	0.1	0.1	0.1	86	43
Montenegro	73	70	75	64	—	—	—	—	39	17
Portugal	78	75	82	55	0.5	0.5	0.5	0.3	67	63
San Marino	82	80	85	84	—	—	—	—	—	—
Serbia	73	71	76	56	0.1[f]	0.1[f]	0.1[f]	0.1[f]	41[f]	19[f]
Slovenia	78	75	82	48	—	<0.1	—	—	74	59
Spain	81	78	84	77	0.5	0.5	0.6	0.2	72	67
OCEANIA	76	74	78	67	0.1	0.4	0.2	0.1	—	59
Australia	81	79	84	83	0.1	0.2	0.2	<0.1	85	75
Federated States of Micronesia	68	67	68	22	—	—	—	—	—	70
Fiji	68	66	71	51	0.1	0.1	0.1	—	—	40
French Polynesia	74	72	77	53	—	—	—	—	—	—
Guam	78	75	82	93	—	—	—	—	—	—
Kiribati	61	59	63	44	—	—	—	—	—	21
Marshall Islands	66	64	67	68	—	—	—	—	45	42
Nauru	56	55	57	100	—	—	—	—	36	25
New Caledonia	76	72	80	58	—	—	—	—	—	—
New Zealand	80	78	82	86	0.1	0.1	0.1	—	—	72
Palau	69	66	72	77	—	—	—	—	—	17
Papua New Guinea	59	57	62	13	0.3	1.5	0.6	0.7	—	9
Samoa	73	72	74	22	—	—	—	—	—	54
Solomon Islands	62	62	63	17	—	—	—	—	—	16
Tonga	71	70	72	24	—	—	—	—	—	23
Tuvalu	64	62	65	47	—	—	—	—	—	29
Vanuatu	67	66	69	21	—	—	—	—	—	15

POPULATION, HEALTH, AND ENVIRONMENT DATA AND ESTIMATES FOR THE COUNTRIES AND REGIONS OF THE WORLD	GNI PPP per Capita (US$) 2008	Population per Square Kilometer	Percent Living on Less Than US$2/Day, 2005	Metric Tons of CO2 Emissions per Capita, 2006
WORLD	$10,090	50	48	4.1
MORE DEVELOPED	32,320	27	—	11.5
LESS DEVELOPED	5,170	67	51	2.4
LESS DEVELOPED (Excl. China)	4,910	58	56	1.8
LEAST DEVELOPED	1,230	40	80	—
AFRICA	2,660	33	65	0.9
SUB-SAHARAN AFRICA	1,950	34	74	0.6
NORTHERN AFRICA	5,370	24	18	1.8
Algeria	7,940	15	24	2.6
Egypt	5,460	79	18	2.1
Libya	15,630	4	—	7.0
Morocco	4,330	71	14	1.3
Sudan	1,930	17	—	0.3
Tunisia	7,070	64	13	1.9
Western Sahara	—	2	—	—
WESTERN AFRICA	1,600	48	76	0.3
Benin	1,460	79	75	0.3
Burkina Faso	1,160	58	81	—
Cape Verde	3,450	126	40	—
Côte d'Ivoire	1,580	66	47	0.3
Gambia	1,280	142	57	—
Ghana	1,430	100	54	0.4
Guinea	1,190	41	87	—
Guinea-Bissau	530	45	78	—
Liberia	300	36	95	—
Mali	1,090	10	77	—
Mauritania	*2,000*	3	44	—
Niger	680	12	86	—
Nigeria	1,940	165	84	0.4
Senegal	1,760	64	60	0.4
Sierra Leone	750	79	76	—
Togo	820	117	69	0.1
EASTERN AFRICA	1,030	49	78	—
Burundi	380	298	93	—
Comoros	1,170	302	65	—
Djibouti	2,330	37	41	—
Eritrea	630	43	—	0.1
Ethiopia	870	75	78	0.1
Kenya	1,580	67	40	0.3
Madagascar	1,040	33	90	—
Malawi	830	120	90	—
Mauritius	12,480	625	—	—
Mayotte	—	503	—	—
Mozambique	770	27	90	0.1
Reunion	—	324	—	—
Rwanda	1,010	375	90	—
Seychelles	19,770	191	—	—
Somalia	—	14	—	—
Tanzania	1,230	46	97	0.1
Uganda	1,140	127	76	—
Zambia	1,230	17	82	0.2
Zimbabwe	—	32	—	0.8
MIDDLE AFRICA	1,650	19	74	0.1
Angola	5,020	14	70	0.5
Cameroon	2,180	40	58	0.2
Central African Republic	730	7	82	—
Chad	1,160	8	83	—
Congo	3,090	11	74	0.3
Congo, Dem. Rep.	290	29	80	z
Equatorial Guinea	21,700	24	—	—
Gabon	12,270	6	20	1.6
Sao Tome and Principe	1,780	169	—	—

See Notes on page 17.

POPULATION, HEALTH, AND ENVIRONMENT DATA AND ESTIMATES FOR THE COUNTRIES AND REGIONS OF THE WORLD

	GNI PPP per Capita (US$) 2008	Population per Square Kilometer	Percent Living on Less Than US$2/Day, 2005	Metric Tons of CO2 Emissions per Capita, 2006
SOUTHERN AFRICA	9,380	22	45	6.3
Botswana	13,100	3	49	2.4
Lesotho	2,000	70	62	—
Namibia	6,270	3	62	1.5
South Africa	9,780	42	43	7.2
Swaziland	5,010	68	81	—
AMERICAS	23,380	22	—	8.4
NORTHERN AMERICA	45,890	16	—	18.4
Canada	36,220	3	—	16.5
United States	46,970	32	—	19.0
LATIN AMERICA/CARIBBEAN	10,160	28	18	2.4
CENTRAL AMERICA	11,900	61	10	3.1
Belize	6,040	14	—	—
Costa Rica	10,950	88	9	1.4
El Salvador	6,670	349	25	0.9
Guatemala	4,690	129	24	0.8
Honduras	3,870	67	35	1.0
Mexico	14,270	56	5	4.0
Nicaragua	2,620	44	32	0.7
Panama	11,650	46	18	1.8
CARIBBEAN	—	176	—	2.2
Antigua and Barbuda	20,570	199	—	—
Bahamas	—	25	—	—
Barbados	—	653	—	—
Cuba	—	101	—	2.4
Dominica	8,300	96	—	—
Dominican Republic	7,890	207	15	1.9
Grenada	8,060	308	—	—
Guadeloupe	—	240	—	—
Haiti	1,180	333	72	0.2
Jamaica	7,360	246	6	4.3
Martinique	—	368	—	—
Netherlands Antilles	—	254	—	20.8
Puerto Rico	—	447	—	—
St. Kitts-Nevis	15,170	191	—	—
Saint Lucia	9,190	319	41	—
St. Vincent & the Grenadines	8,770	283	—	—
Trinidad and Tobago	23,950	260	14	20.3
SOUTH AMERICA	10,160	22	19	2.2
Argentina	14,020	14	11	3.8
Bolivia	4,140	9	30	1.4
Brazil	10,070	22	18	1.8
Chile	13,270	22	5	3.6
Colombia	8,510	40	26	1.3
Ecuador	7,760	48	20	1.9
French Guiana	—	3	—	—
Guyana	2,510	4	17	—
Paraguay	4,820	16	18	0.6
Peru	7,980	23	19	1.0
Suriname	7,130	3	27	—
Uruguay	12,540	19	5	1.9
Venezuela	12,830	31	32	5.5
ASIA	$6,020	129	53	3.0
ASIA (Excl. China)	6,020	125	62	2.4
WESTERN ASIA	10,720	48	—	5.3
Armenia	6,310	104	43	1.4
Azerbaijan	7,770	101	<2	3.6
Bahrain	—	1,754	—	27.0
Cyprus	*24,040*	116	—	9.1
Georgia	4,850	66	30	1.0
Iraq	—	69	—	3.2

See Notes on page 17.

POPULATION, HEALTH, AND ENVIRONMENT DATA AND ESTIMATES FOR THE COUNTRIES AND REGIONS OF THE WORLD

	GNI PPP per Capita (US$) 2008	Population per Square Kilometer	Percent Living on Less Than US$2/Day, 2005	Metric Tons of CO2 Emissions per Capita, 2006
Israel	27,450	345	—	8.9
Jordan	5,530	66	4	3.3
Kuwait	*52,610*	168	—	25.7
Lebanon	10,880	373	—	3.3
Oman	*20,650*	10	—	11.6
Palestinian Territory	—	653	—	—
Qatar	—	128	—	48.3
Saudi Arabia	*22,950*	13	—	14.4
Syria	4,350	118	—	2.7
Turkey	13,770	95	9	3.3
United Arab Emirates	—	61	—	26.0
Yemen	2,210	43	47	0.9
SOUTH CENTRAL ASIA	3,120	160	70	1.3
Afghanistan	—	44	—	—
Bangladesh	1,440	1,127	81	0.2
Bhutan	4,880	15	50	—
India	2,960	356	76	1.1
Iran	*10,840*	44	8	6.2
Kazakhstan	9,690	6	17	11.9
Kyrgyzstan	2,130	27	52	1.0
Maldives	5,280	1,057	—	—
Nepal	1,120	187	78	0.1
Pakistan	2,700	227	60	0.8
Sri Lanka	4,480	312	40	0.6
Tajikistan	1,860	52	51	0.9
Turkmenistan	6,210	10	50	8.9
Uzbekistan	2,660	62	77	4.3
SOUTHEAST ASIA	4,520	133	44	1.6
Brunei	*50,200*	66	—	15.1
Cambodia	1,820	82	68	0.3
Indonesia	3,830	128	54	1.5
Laos	2,060	27	77	—
Malaysia	13,740	86	8	5.9
Myanmar	*1,290*	74	—	0.2
Philippines	3,900	307	45	0.8
Singapore	47,940	7,486	—	9.6
Thailand	5,990	132	12	3.4
Timor-Leste	4,690	76	78	—
Vietnam	2,700	263	48	1.0
EAST ASIA	9,100	133	36	5.0
China	6,020	139	36	4.3
China, Hong Kong SAR[c]	43,960	6,403	—	6.1
China, Macao SAR[c]	*52,260*	21,346	—	—
Japan	35,220	338	—	9.5
Korea, North	—	188	—	3.2
Korea, South	28,120	490	—	9.9
Mongolia	3,480	2	49	3.9
Taiwan	—	641	—	11.9
EUROPE	25,550	32	—	8.4
NORTHERN EUROPE	36,090	55	—	8.5
Channel Islands	—	804	—	—
Denmark	37,280	128	—	10.2
Estonia	19,280	30	<2	11.3
Finland	35,660	16	—	12.7
Iceland	25,220	3	—	7.2
Ireland	37,350	64	—	10.6
Latvia	16,740	35	<2	3.5
Lithuania	18,210	51	<2	4.0
Norway	58,500	13	—	7.9
Sweden	38,180	21	—	5.3
United Kingdom	36,130	255	—	8.9

POPULATION, HEALTH, AND ENVIRONMENT DATA AND ESTIMATES FOR THE COUNTRIES AND REGIONS OF THE WORLD

	GNI PPP per Capita (US$) 2008	Population per Square Kilometer	Percent Living on Less Than US$2/Day, 2005	Metric Tons of CO2 Emissions per Capita, 2006
WESTERN EUROPE	36,430	170	—	8.7
Austria	37,680	100	—	8.8
Belgium	34,760	354	—	11.1
France	34,400	114	—	6.0
Germany	35,940	230	—	10.0
Liechtenstein	—	223	—	—
Luxembourg	64,320	193	—	23.6
Monaco	—	35,382	—	—
Netherlands	41,670	398	—	10.9
Switzerland	46,460	188	—	5.8
EASTERN EUROPE	14,430	16	2	8.9
Belarus	12,150	47	<2	6.6
Bulgaria	11,950	68	<2	6.2
Czech Republic	22,790	133	<2	11.8
Hungary	17,790	108	<2	5.6
Moldova	3,210	122	29	1.9
Poland	17,310	122	<2	8.0
Romania	13,500	90	3	4.4
Russia	15,630	8	<2	11.1
Slovakia	21,300	110	<2	7.0
Ukraine	7,210	76	<2	6.6
SOUTHERN EUROPE	26,740	118	—	7.0
Albania	7,950	111	8	1.3
Andorra	—	184	—	—
Bosnia-Herzegovina	8,620	75	<2	4.4
Croatia	18,420	78	<2	4.7
Greece	28,470	85	—	8.4
Italy	30,250	200	—	7.6
Kosovo[d]	—	204	—	—
Macedonia[e]	9,950	80	3	3.9
Malta	*22,460*	1,310	—	6.2
Montenegro	13,920	45	—	—
Portugal	22,080	116	—	5.3
San Marino	—	515	—	—
Serbia	11,150	95	—	7.2
Slovenia	26,910	101	<2	7.7
Spain	31,130	93	—	7.4
OCEANIA	24,360	4	—	—
Australia	34,040	3	—	19.0
Federated States of Micronesia	3,000	158	—	—
Fiji	4,270	46	—	—
French Polynesia	—	67	—	—
Guam	—	332	—	—
Kiribati	3,660	136	—	—
Marshall Islands	—	298	—	—
Nauru	—	465	—	—
New Caledonia	—	14	—	—
New Zealand	25,090	16	—	8.9
Palau	—	45	—	—
Papua New Guinea	2,000	14	57	—
Samoa	4,340	67	—	—
Solomon Islands	2,580	18	—	—
Tonga	3,880	138	—	—
Tuvalu	—	427	—	—
Vanuatu	3,940	20	—	—

NOTES

(—) Indicates data unavailable or inapplicable.

a Infant deaths per 1,000 live births. Rates shown with decimals indicate national statistics reported as completely registered, while those without are estimates from the sources cited on reverse. Rates shown in italics are based upon fewer than 50 annual infant deaths and, as a result, are subject to considerable yearly variability.

b Average number of children born to a woman during her lifetime.

c Special Administrative Region.

d Kosovo declared independence from Serbia on Feb. 17, 2008. Serbia has not recognized Kosovo's independence.

e The former Yugoslav Republic.

f Includes Kosovo.

z Rounds to zero.

* Data prior to 2003 are shown in italics.

For additional notes and sources, see reverse side.

Data prepared by PRB demographers Carl Haub and Mary Mederios Kent.

Acknowledgments, Notes, Sources, and Definitions

ACKNOWLEDGMENTS

The authors gratefully acknowledge the valuable assistance of PRB staff members Donna Clifton, Jennay Ghowrwal, and Kelvin Pollard; staff of the International Programs Center of the U.S. Census Bureau; the United Nations (UN) Population Division; the Institut national d'etudes démographiques (INED), Paris; and the World Bank in the preparation of this year's *World Population Data Sheet*. This publication is funded by the William and Flora Hewlett Foundation, the David and Lucile Packard Foundation, the U.S. Agency for International Development (Cooperative Agreement GPO-A-00-03-00004-00), and supporters. The information in this *Data Sheet* was not provided by and does not represent the views of the United States government or the U.S. Agency for International Development.

NOTES

The *Data Sheet* lists all geopolitical entities with populations of 150,000 or more and all members of the UN. These include sovereign states, dependencies, overseas departments, and some territories whose status or boundaries may be undetermined or in dispute. **More developed** regions, following the UN classification, comprise all of Europe and North America, plus Australia, Japan, and New Zealand. All other regions and countries are classified as **less developed**. The **least developed** countries consist of 49 countries with especially low incomes, high economic vulnerability, and poor human development indicators. The criteria and list of countries, as defined by the United Nations, can be found at www.unohrlls.org/en/ldc/.

Sub-Saharan Africa: All countries of Africa except the northern African countries of Algeria, Egypt, Libya, Morocco, Tunisia, and Western Sahara.

World and Regional Totals: Regional population totals are independently rounded and include small countries or areas not shown. Regional and world rates and percentages are weighted averages of countries for which data are available; regional averages are shown when data or estimates are available for at least three-quarters of the region's population.

World Population Data Sheets from different years should **not be used as a time series**. Fluctuations in values from year to year often reflect revisions based on new data or estimates rather than actual changes in levels. Additional information on likely trends and consistent time series can be obtained from PRB, and are also available in UN and U.S. Census Bureau publications and websites.

SOURCES

The rates and figures are primarily compiled from the following sources: official country statistical yearbooks, bulletins, and websites; *United Nations Demographic Yearbook, 2006* of the UN Statistics Division; *World Population Prospects: The 2008 Revision* of the UN Population Division; and the International Data Base of the International Programs Center, U.S. Census Bureau. Other sources include recent demographic surveys such as the Demographic and Health Surveys, Reproductive Health Surveys, special studies, and direct communication with demographers and statistical bureaus in the United States and abroad. Specific data sources may be obtained by contacting the authors of the *2009 World Population Data Sheet*.

For countries with complete registration of births and deaths, rates are those most recently reported. For more developed countries, nearly all vital rates refer to 2007 or 2008.

DEFINITIONS

Mid-2009 Population

Estimates are based on a recent census, official national data, or PRB, UN, and U.S. Census Bureau projections. The effects of refugee movements, large numbers of foreign workers, and population shifts due to contemporary political events are taken into account to the extent possible.

Birth and Death Rate

The annual number of births and deaths per 1,000 total population. These rates are often referred to as "crude rates" since they do not take a population's age structure into account. Thus, crude death rates in more developed countries, with a relatively large proportion of high-mortality older population, are often higher than those in less developed countries with lower life expectancy.

Rate of Natural Increase (RNI)

The birth rate minus the death rate, implying the annual rate of population growth without regard for migration. Expressed as a percentage.

Net Migration

The estimated rate of net immigration (immigration minus emigration) per 1,000 population for a recent year based upon the official national rate or derived as a residual from estimated birth, death, and population growth rates. Migration rates can vary substantially from year to year for any particular country as can the definition of an immigrant.

Projected Population 2025 and 2050

Projected populations based upon reasonable assumptions on the future course of fertility, mortality, and migration. Projections are based upon official country projections, series issued by the UN or the U.S. Census Bureau, or PRB projections.

Infant Mortality Rate

The annual number of deaths of infants under age 1 per 1,000 live births. Rates shown with decimals indicate national statistics reported as completely registered, while those without are estimates from the sources cited above. Rates shown in italics are based upon fewer than 50 annual infant deaths and, as a result, are subject to considerable yearly variability.

Total Fertility Rate (TFR)

The average number of children a woman would have assuming that current age-specific birth rates remain constant throughout her childbearing years (usually considered to be ages 15 to 49).

Population Under Age 15/Age 65+

The percentage of the total population in these ages, which are often considered the "dependent ages."

Life Expectancy at Birth

The average number of years a newborn infant can expect to live under current mortality levels.

Percent Urban

Percentage of the total population living in areas termed "urban" by that country. Countries define urban in many different ways, from population centers of 100 or more dwellings to only the population living in national and provincial capitals.

Prevalence of HIV/AIDS

The estimated percentage of adults, ages 15-49, and youth, ages 15-24, living with HIV/AIDS. Most figures are from UNAIDS' *2008 Report on the Global AIDS Epidemic*. Figures are updated with data from Demographic and Health Surveys where available. Some regional averages were calculated by PRB.

Contraceptive Use

The percentage of currently married or "in-union" women of reproductive age who are currently using any form of contraception. "Modern" methods include clinic and supply methods such as the pill, IUD, condom, and sterilization. Data are from the most recently available national-level surveys, such as Demographic and Health Surveys, Reproductive Health Surveys, Multiple Indicator Cluster Surveys, regional survey programs, national surveys, and the UN Population Division *World Contraceptive Use 2007*. For more developed countries, data refer to some point in the 1990s and early 2000s; and for less developed countries, from 1995. Data prior to 2003 are shown in italics.

GNI PPP per Capita, 2008 (US$)

GNI PPP per capita is gross national income in purchasing power parity (PPP) divided by midyear population. GNI PPP refers to gross national income converted to "international" dollars using a purchasing power parity conversion factor. International dollars indicate the amount of goods and services one could buy in the United States with a given amount of money. Data are from the World Bank. Figures in italics are for 2006 or 2007.

Percent of Population Living on Less Than US$2/Day

The proportion of the population living on less than US$2 per day at 2005 international prices, which are prices converted using purchasing power parity (PPP) rates. PPP is the amount of a country's currency required to buy the same amount of goods and services in the country's domestic market as a U.S. dollar would buy in the United States. The World Bank's estimates are drawn from surveys that use common methods for measuring household living standards across countries. Most estimates refer to the 2001-2007 period. Figures are taken from the World Bank, *2008 World Development Indicators, Poverty Data, A Supplement to World Development Indicators 2008*.

Carbon Dioxide Emissions per Capita

The amount of CO_2 in metric tons produced or emitted per person. Figures shown include carbon dioxide produced through fuel combustion. Figures are taken from the International Energy Agency, *Key World Energy Statistics 2008*.

For a full list of PRB publications available in English, French, Spanish, Arabic, and Portuguese, go to PRB's online store at www.prb.org.

To order PRB publications (discounts available for bulk orders):

- Online at www.prb.org.
- E-mail: popref@prb.org.
- Call toll-free: 800-877-9881.
- Fax: 202-328-3937.
- Mail: 1875 Connecticut Ave., NW, Suite 520, Washington, DC 20009.

The *2009 World Population Data Sheet* is also available in French and Spanish.

Data prepared by PRB demographers Carl Haub and Mary Mederios Kent.

Design and production: Michelle Corbett, Black Mountain Creative.

ISSN 0085-8315

Photo Credits from cover, top to bottom: © 2005 Omar Mohsen, Courtesy of Photoshare; © 2006 Richard Nyberg, Courtesy of Photoshare; © 2007 naphtalina, iStockPhoto; © 2007 Alfredo L. Fort, Courtesy of Photoshare

The **POPULATION REFERENCE BUREAU informs** people around the world about population, health, and the environment, and **empowers** them to use that information to **advance** the well-being of current and future generations.

INFORM

PRB informs people around the world and in the United States about issues related to population, health, and the environment. To do this, we transform technical data and research into accurate, easy-to-understand information.

INNOVATIVE TOOLS. PRB's wallcharts, including the *World Population Data Sheet* and *Family Planning Worldwide 2008*, are searchable via our DataFinder web tool and make accurate demographic information accessible to a wide audience.

INFLUENTIAL REPORTS. *Female Genital Mutilation/Cutting: Data and Trends* is a new and unique data sheet that highlights FGM/C data and information for 27 countries. This data sheet is proving to be useful to those working for the abandonment of FGM/C and for the improvement of women's reproductive health.

UNBIASED POLICY ANALYSIS. For more than 20 years, PRB has hosted a monthly seminar series focused on the policy implications of population issues including the U.S. Census 2010 and climate change and urban adaptation.

ONLINE RESOURCES. PRB's website offers full text of all PRB publications, including our respected *Population Bulletins* and web-exclusive data and analysis on world issues ranging from aging to family planning. Our Center for Public Information on Population Research puts new population research findings into context for journalists and policymakers.

EMPOWER

PRB empowers people—researchers, journalists, policymakers, and educators—to use information about population, health, and the environment to encourage action.

Information alone can be powerful. Frequently, however, people have knowledge but lack the tools needed to communicate it effectively to decisionmakers. PRB builds coalitions and conducts trainings in the United States and throughout the developing world to share techniques to inform policy.

JOURNALIST NETWORKS. Since 1996, PRB has shared techniques for fact-based, reproductive health reporting with a network of West African editors. The Pop'Médiafrique program, one of several PRB journalist networks, has improved news coverage and increased demand for family planning in the region.

POLICY COMMUNICATIONS TRAINING. PRB has trained more than 700 professionals in Asia, Africa, and Latin America. For example, PRB collaborated with the West Africa Health Organization, WHO/AFRO, and AED to train 36 professionals from eight West African countries in how to advocate for strengthening family planning services to high-level policymakers and opinion leaders.

DATA WORKSHOPS. PRB's data workshops assist the Annie E. Casey Foundation's KIDS COUNT network in using vital data about the status of children in the United States. Workshop participants take away the knowledge needed to access data about their particular state and communicate with policymakers.

ADVANCE

PRB works to advance the well-being of current and future generations. Toward that end, we analyze data and research, disseminate information, and empower people to use that information in order to inform policymaking.

While the numbers of publications created or workshops conducted are one way to measure PRB's work, the creation of evidence-based policies, increased demand for health services, and active coalitions are better gauges of progress toward positive social change.

EVIDENCE-BASED POLICIES. PRB provides analysis for the KIDS COUNT Data Book, an annual report card on the well-being of children and families in the United States, that has helped promote the passage of several U.S. policies, including the State Children's Health Insurance Program.

DEMAND FOR HEALTH SERVICES. Information broadcast by women radio journalists who attended PRB's reproductive health workshop in Senegal has increased local demand for family planning and health services.

ACTIVE COALITIONS. In March 2008, PRB collaborated with Filipino stakeholders in hosting the Third National Population, Health, and Environment Conference, which brought together 326 national and international representatives from government, civil society, and the private sector. This conference provided a forum for sharing experiences and best practices.

PRB's *World Population Data Sheet* is used around the world and is widely considered to be the most accurate source of information on population. If you value the *Data Sheet* and are among the tens of thousands of people who eagerly anticipate its publication each year, please consider making a contribution to PRB. Your donation will help ensure that PRB can maintain its commitment to keeping the *Data Sheet* as affordable as possible. Visit our website to donate now: www.prb.org.

PRB | INFORM EMPOWER ADVANCE

USAID
FROM THE AMERICAN PEOPLE

1875 Connecticut Ave., NW, Washington, DC 20009 USA
tel. 202-483-1100 | fax 202-328-3937 | e-mail: popref@prb.org | website: www.prb.org